Transformed cladistics,
taxonomy and evolution

Transformed cladistics, taxonomy and evolution

N. R. SCOTT-RAM

Biotechnology Consultant

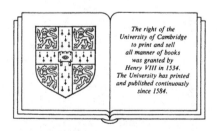

The right of the
University of Cambridge
to print and sell
all manner of books
was granted by
Henry VIII in 1534.
The University has printed
and published continuously
since 1584.

CAMBRIDGE UNIVERSITY PRESS

Cambridge

New York   Port Chester

Melbourne   Sydney

CAMBRIDGE UNIVERSITY PRESS
Cambridge, New York, Melbourne, Madrid, Cape Town, Singapore, São Paulo

Cambridge University Press
The Edinburgh Building, Cambridge CB2 8RU, UK

Published in the United States of America by Cambridge University Press, New York

www.cambridge.org
Information on this title: www.cambridge.org/9780521340861

© Cambridge University Press 1990

First published 1990
This digitally printed version 2008

A catalogue record for this publication is available from the British Library

Library of Congress Cataloguing in Publication data
Scott-Ram, N. R.
    Transformed cladistics, taxonomy, and evolution / N. R. Scott-Ram.
        p.   cm.
    Based on the author's thesis (Ph.D.).
    Bibliography: p.
    Includes index.
    ISBN 0 521 34086 1
    1. Biology–Classification. 2. Cladistic analysis. 3. Evolution.
    I. Title.
    QH83.S36   1989
    574'.012–dc20   89–9754   CIP

ISBN 978-0-521-34086-1 hardback
ISBN 978-0-521-05513-0 paperback

*To my mother and father
and uncle Jerry*

CONTENTS

PREFACE

A sound understanding of the interrelationship between evolutionary theory and taxonomic theory is probably more important now than it ever has been since Darwin first developed his theory of evolution in *The Origin of Species by Means of Natural Selection*. This is not just because there are doubts over the precise nature of this relationship, as exemplified in the debate on pattern and process in taxonomy, but also because there are advances being made in other biological disciplines, such as molecular genetics, immunology, protein chemistry and microbiology, which call into question the limits of our understanding of structure and function in biological systems.

Since Darwin, the hierarchy, as discovered in nature through classification, has been wedded to evolutionary theory. Evolution provided an explanation which accounted for our perception of the hierarchy. In the past decade, this view has come under fire from a number of quarters, most conspicuously from taxonomists themselves. Surprisingly this attack has not been grounded in recent advances from other empirical disciplines: rather, taxonomists have argued over what the 'proper' methods of classification should be, generally motivated by a desire to find some adamantine system which is not at the mercy of theoretical change. The situation has some similarities with early twentieth-century physics, where in the face of a rapidly advancing theoretical science, retrenchment to a position of dogmatism by opponents of the advances was never very far away. In the taxonomic debate today, what is perhaps most strange is the obsession with questions of epistemology, methodology and ontology, and the poor attempts to hijack outdated perspectives from the philosophy of science. My principal aim in this book will be to show that the debates in philosophy of science to which various taxonomic schools of thought appeal, are irrelevant and misguided.

It is not my aim to discuss either that evolutionary theory is in a state of disarray as has been suggested or that it is in a healthy state of growth fuelled by rapidly emerging empirical advances. Rather, my aim is to uncover underlying issues in taxonomy from the last twenty years and compare the various schools of thought in terms of their own respective aims and criteria of what 'good' taxonomy is.

At the same time, I aim to document two basic approaches in taxonomy which reveal hidden attitudes about the status of evolutionary theory, and more generally, about the status of theories and their relationship to taxonomic practice. Whatever the advantages and disadvantages of particular viewpoints aired in taxonomy during the last twenty years, what should be apparent, but is often ignored, is that taxonomic practice cannot become divorced from underlying empirical advances from related disciplines. A taxonomy divorced from empirical data is likely to be of little practical or theoretical use.

In a book of this length it is difficult to do justice to all of the topics involved in taxonomy. In particular, I do not dwell at any great length on the burgeoning discipline of computer modelling in cladistic analysis. Nor has a great deal of attention been paid to issues such as whether species are individual or classes, or to other philosophical issues that are now being raised within biology in general. In these deficiences, I ask for the reader's tolerance.

My intellectual and psychological debts are diverse. In the former, two people deserve special thanks: Adrian Friday for first arousing my interest in phylogenetic reconstruction, and the wider issues relating to evolutionary theory. His assistance and enthusiasm for this work were a constant source of support and greatly appreciated. The other debt is to Nick Jardine, whose shrewdness and depth of understanding helped clarify many underlying issues. The intellectual debt I owe him will always remain with me.

Thanks must also go to Christine Janis who provided both moral and intellectual support to this enterprise. Her working knowledge of many paleontological issues was of invaluable help, as were her 'Rubber Chicken' Seminars. I also owe a special debt to Alec Panchen, who encouraged me to write this book on the basis of my Ph.D thesis. Others from who I have benefited in discussion and general assistance include Martin Bishop, Jenny Clark, Beverley Halstead and Chris Humphries.

On the more psychological side of things, several people deserve special mention for putting up with me during the gestation of this book; Tony Morris for supporting me during low points. Likewise Fiona Lennon gave me continued encouragement and support during my endeavours. Yaron

Meshoulam kindly provided word processing facilities which made preparation of early drafts of this manuscript much easier than otherwise could have been. I also owe a great debt to Andra Houchen who typed up the final drafts, and without whom circumstances would have made it difficult to finish this work. Finally, I would like to thank two people who, while not being directly involved in this work have provided invaluable support in a number of subtler ways – Derek Lennon and Eric Crook.

INTRODUCTION

For over twenty years the world of taxonomy has been riddled with controversy. This might be interpreted as a healthy state of affairs in any scientific discipline: but it is only recently that the debate has had significant repercussions for Neo-Darwinism in general. Doubt has been cast on the view that the history of life, as indicated by the fossil record, is consistent with the evolutionary process of genetic mutation, guided towards the adaptation of populations by natural selection. More specifically, a small group of taxonomists, known as the 'transformed cladists', have questioned Darwin's fundamental assumption in *The Origin* that the natural hierarchy in nature was evidence for evolution. Claims that Darwin got it wrong have abounded,[1] and have been enhanced by findings from other areas such as developmental biology, macroevolutionary theory, and analysis of speciation models.

For the Neo-Darwinian, insult has been added to injury by the re-emergence, especially in the USA, of creationist doctrines, which attempt to challenge evolutionary theory on scientific grounds,[2] interpreting well-established scientific facts in a manner that is consistent with creationist thinking. Not surprisingly, admist all this turmoil, accuracy of fact and argument have been traded off against sensationalist media coverage.[3]

Some have welcomed taxonomy's entry into the limelight, others have been horrified at the prospect of the disappearance of long cherished traditions, such as the role of paleontology in evolutionary studies. The status of evolutionary theory is apparently at stake. Clearly there is a need to separate picturesque distortion from hard analysis, to reappraise the arguments raised in the taxonic controversy, and to examine the implications for Neo-Darwinism. Taxonomy's methodological disarray over the past two decades makes this enterprise all the more pertinent. For this reason, some attention must be paid to historical detail. But before outlining the historical development of the contemporary schools of taxonomy, the precise boundaries of inquiry must be delineated.

The boundaries of inquiry

Several preliminary definitions are relevant to the discussion as a whole. The biological discipline known as systematics deals with the theory and practice of capturing orderliness in nature. Systematics may be defined, following Simpson, as

> the scientific study of all kinds and diversity of organisms and of any and all relationships among them.[4]

Taxonomy, on the other hand, has a more restricted meaning and is concerned with the theory and practice of classifying organisms. Explicity,

> Taxonomy is the theoretical study of classification, including its bases, principles, procedures, and rules.[5]

There is some overlap between taxonomy and classification in that the activity of classification is consistent with taxonomy, given a basic definition of classification as the ordering of organisms, both living and dead, into groups (or sets) on the basis of their similarities and relationships. But even this definition of classification requires some clarification. Hull has suggested three meanings for the words *a classification*;

> At one extreme, a classification is nothing but a list of taxa names indented to indicate category levels. Others would also include all the characters and the taxonomic principles used to construct a classification as part of the classification. At the other extreme, some authors use the words *a classification* to refer to the entire taxonomic monograph.[6]

Hull's point is that when we use the term 'classification', two modes of use are possible – firstly, in referring to the act of classifying, which involves the ordering of organisms into groups, and secondly, in referring to the finished product of the classifying activity, the classification. For the most part, we will be concerned primarily with the former sense, but when reference is made to the classification as a finished product, only the restricted definition of a classification, as a list of taxa names indented to indicate category levels, will be employed.

In addition to this process–product distinction, there is a further dichotomy within the process of classification, between the formation of a system of classes and the process of assigning additional, unidentified objects to these classes. The latter is more commonly known as diagnosis or identification in contradistinction to the process of classification.

The third distinction dates back to the modern synthesis of the 1940s, in which questions concerning classification centred on two different aspects – the 'species problem', and how such species were to be classified into higher taxa such as genera, families and orders. These two aspects survive today in the fields of microtaxonomy and macrotaxonomy.

The species problem is tackled in microtaxonomy and this concerns 'the delimitation of species with the analysis of the genetical and ecological barriers between populations and with the investigation of the ranges of morphological and genetic variability of populations in relation to their environments.'[7] In addition to delimiting taxa of specific rank, microtaxonomy confronts the problem of classifying groupings of populations at the infraspecific level. Of direct relevance here are the results of experimental taxonomy, which focuses on patterns of variation at the infraspecific level.[8] Microtaxonomy, therefore, embraces a wide range of views and this is in sharp contrast to macrotaxonomy, which deals specifically with the grouping of species into higher taxa. In this book I will be concerned with macro-taxonomy alone.

It is clear that macrotaxonomy is likely to be sensitive to the results of microtaxonomy, and while this is certainly apparent in the direct relevance of speciation modes to macrotaxonomy, there is some evidence that the contact between the two fields is not as close as it might be. Indeed, the results of macrotaxonomy appear to be of little relevance to microtaxonomy, indicating that the disputes in macrotaxonomy are too divorced from an experimental basis. Certainly this is true of all cladists' dismissal of the results of population genetics.[9]

Historical order of appearance of the schools

Four contemporary schools of taxonomy are generally recognised: evolutionary systematics, phenetics, phylogenetic cladistics, and transformed cladistics. This diversity of opinion can be traced back to two phases of development in the continuing debate over the proper methods and foundations of biological classification. During each phase, the emergence of a new school of thought increased the urgency of the various disputes by clarifying some of the important issues at stake. In the late 1960s the first phase was initiated by the arrival of the phenetic school of taxonomy, as typified in Sokal & Sneath's *Principles of Numerical Taxonomy*,[10] which concretised issues over the relationship between classification and phylogeny (the reconstructed history of life). Pheneticists advocated the grouping of organisms together on the basis of their similarity, whilst rejecting the use of phylogenetic information in the construction of such groups. The increasing use of computers also added force to the desire to explicate and quantify taxonomy than had hitherto been done before. Classics such as Simpson's *Principles of Animal Taxonomy*,[11] and Mayr's *Principles of Systematic Zoology*,[12] which outlined the field of inquiry of the evolutionary systematists, typically came under fire. For their part, the evolutionary systematists argued for a taxonomy that represented groups of organisms which were a product of evolution, and that classifications, therefore, should be based upon all aspects of phylogeny.

During this period, the English translation of Hennig's *Phylogenetic Systematics*[13] was picked upon by a few devotees, e.g. Brundin,[14] and through such commentaries grew in stature. Hennig's book formed the basis for cladistics, which advocated a taxonomy that mirrored evolutionary branching sequences. Classifications were based on genealogy (relationships of descent) alone. Such views gained acceptance by the early 1970s, and set the stage for the second phase of thinking – an examination of the precise relationship between classifications and evolutionary theory. The focal point was a questioning of the validity of evolutionary assumptions in the procedures of classification. This problem was highlighted in the split between the phylogenetic cladists, who rigidly adhered to the basic postulates of Hennig, and the transformed cladists, who attempted to pick out the kinds of patterns that Hennig viewed as evidence for evolution, whilst abandoning interest in the branching sequence itself. The recently emerged transformed cladists, led by Nelson, [15] Patterson [16] and Platnick,[17] not only dropped the evolutionary portion from their classifications, but also hinted at an anti-Darwinian element in suggesting a return to pre-Darwinian classifications based on recurrent patterns in nature. These claims meant that the debate in taxonomy took a radical turn[18] because not only were the use of evolutionary assumptions in classification called into question, but also Darwin's point, concerning the evidential priority of the natural hierarchy in evolutionary theory, was highlighted.

Both phenetics and phylogenetic cladistics have undergone substantial modifications in response to the sometimes cumbersome attempts of the fourth, and oldest school – the evolutionary systematists – to stave off their polemical attacks. These changes have not always been very explicit, with frequent arguments concentrating on questions of terminology. Nowhere is this greater than in the shift from phylogenetic cladistics to transformed cladistics. Often the arguments have grown out of all proportion to their actual content, so that the underlying issues have remained shrouded in terminological confusion. The diversity of opinion and difficulty in defining a common ground has often meant that 'definitive' definitions and straw men abound. If these tendencies are to be avoided a sympathetic and realistic account of the issues involved must be given, with a view to confronting problems that are relevant to all schools of thought.

Relevant issues

The issues that will be confronted in this book can be put under four general headings

(i) What are the aims and purposes of classifications?

(ii) Are these aims sensible? Are they worth pursuing? Emphasis will be placed on the justification of these aims and purposes.

(iii) What methods do we use in constructing a classification?

(iv) Are such methods concordant with the aims? Do the techniques employed actually achieve what they are meant to? What are the limits of these aims and methods?

Any careful comparison of the issues raised in the 1960s and the 1970s will show up similarities, implying that the debate in the 1970s is a re-run of some of the problems raised a decade earlier. This is especially the case when questions concerning the storage of information in classifications and the removal of bias in classification construction, are confronted. Both the pheneticists and transformed cladists advocate the removal of bias, while all four schools claim that their respective classifications maximise the storage and retrieval of information. In considering the limits of each school's aims and methods, inconsistencies in argument are apparent, and the resulting conclusions can help to answer questions concerning the validity of such aims. In this way, each school is examined with respect to its *own* standards and norms, thereby avoiding over simplification. Discussion of the problems faced by all four schools will constitute the main body of the book, and emphasis will be placed on possible reasons for the shifts in taxonomy. In the case of transformed cladistics, and its possible relevance to Neo-Darwinism, some authors[19] have claimed that current research from the history and philosophy of science has prompted this situation. This is incorrect. The current controversy in taxonomy, and evolutionary thinking in general, is not due to philosophy, but stems from specific problems within these areas.

I will begin my account in Chapter 1 with a general discussion of a fundamental difference of approach between the four taxonomic schools. This difference is best explicated through the distinction between descriptive and theoretical 'attitudes' in taxonomy. The former attempts to do away with theoretical assumptions in classification, while the latter relies heavily on theoretical inference. There is a limit to the degree to which these attitudes can be pursued, and at each extreme, philosophical problems are encountered. Part II explores the problems encountered by the evolutionary systematists (Chapter 2) and phylogenetic cladists (Chapter 3). In both cases, the problems are examined in the context of information content in classifications, and their shortcomings are discussed in terms of the theoretical attitude. In Part III, the limits to the descriptive attitude are explored. Thus, in Chapter 4, phenetics is examined in the context of the descriptive attitude, and limits to the elimination of bias in classification construction are explored. The discussion of transformed cladistics in Chapter 5 emphasises similarities and dissimilarities with phenetics, and concludes that the transformed cladists inadvertently incorporate a Platonic world view. In the final chapter, the problem of removing evolutionary assumptions on the basis of methodological considerations alone is examined, and it is argued that if the transformed cladists wish to avoid Platonic assumptions, then their methods are only intelligible in the light of evolutionary theory. I conclude that transsound, and it is doubtful whether it directly challenges the present view of evolution since it is still susceptible to change in evolutionary thinking itself.

A final word of explanation is necessary over the structuring of my account. While the distinction between the theoretical and descriptive attitudes is described in Part I, and then used as a basis for examining the merits of each taxonomic school, the argument in this book should not be construed as *circular* in any way, since the basis for drawing the two attitudes is rooted in the general descriptions of the activities of each school given in the respective chapters.

PART I

Issues pertaining to the
philosophy of science

1

Theoretical and descriptive attitudes in taxonomy

Introduction

Over the years, the taxonomic schools have encountered several recurrent problems. Both the evolutionary systematists and phylogenetic cladists confronted problems over the accurate storage and retrieval of information in classifications, while the pheneticists and transformed cladists concentrated on ways of reducing bias in classification construction. These problems relate to the limits of the aims and methods of each school, and can be situated in a wider domain of discussion, some of which has relevance to the philosophy of science. An important distinction in this respect concerns internal and external issues.

Internal issues

These concern questions that have arisen directly from the debate in taxonomy over the past twenty years, and are primarily concerned with the limits in the aims and procedures of classification. In pursuing these limits, several philosophical problems are raised with respect to the following questions.

(A)　*How much theoretical information can classifications carry?*
This refers specifically to the finished product, a classification.

(B)　*How much theory can be eliminated in the construction of classifications?*
This concerns the elimination of theoretical presuppositions indirectly involved in application to classifications, e.g. the results of comparative morphology. The overall emphasis is on the way in which a classification is constructed with regard to the unloading of theory in any given methodology. In other

words, how much is classification construction the servant of some theory? How far can bias be eliminated from a given method?

(C)   *How is theory involved in the interpretation of a classification?*

Central to this question is the relation between a classification and the underlying theory, and whether or not classifications always require explanatory appeal to a particular theory.

Each of these questions can be expressed in terms of *limits*. The corresponding questions are

(A1) At the extremes are classifications theories on the one hand, or are they empirical devices on the other? As it stands, question (A) is largely an empirical issue relating to the types of relationships between taxa and their expression through indicators of genetics. But, at the extremes, philosophical considerations enter in.

(B1) What are the limits to the elimination of bias? Is all theoretical bias eliminable? In attempting to eliminate bias, we can refer to approaches that use theory or avoid it; in avoiding theory we can talk of the coding of characters or the non-theoretical criterion of homology.

(C1) The extreme case here is whether we can reject the explanatory element in classifications and abandon the search for hidden causes. Is it possible to abandon all recourse to an explanatory element?

External issues

These relate to problems that have not arisen directly out of taxonomy and, typically, concern an appeal to debates that have been occurring within the philosophy of science in the last thirty years. Usually a philosophical stance is made, which is then used to justify a particular position in taxonomy to the detriment of various opponents. Thus the justification for a particular school in taxonomy is made on the strength of philosophical attitudes which are in vogue at any given time. For example, the early pheneticists adopted the views of logical empiricism[1] and operationalism[2] to justify the purging of any metaphysical elements in classification construction. More recently, many cladists[3] have adopted Popper's views on falsification in the elimination of competing hypotheses. The advantage of such strategies is that certain types of taxonomic procedure can be ruled out, while bolstering up one's own position. Of course, not all justifications for a particular view of classification are based on the philosophy of science, but in cases where this occurs, it will be clear that such justifications are misapplied, and often lead to undesirable consequences. Indeed, those issues which have been drawn from the philosophy of science are largely a 'red-herring', and, if anything, have led to inconsistencies in aims and methods of the approaches that utilise such justifications.

In this chapter, we will be concerned primarily with laying the groundwork on the internal issues, since the external issues are best examined in the specific cases where an underlying philosophical view is expressed. In questions (A1), (B1) and (C1), philosophy can help to supply some answers, which in turn can be applied to the unravelling of some underlying problems in taxonomy today.

That the above questions are related is obvious since in arguing for the abandonment of hidden causes, it is a necessary prerequisite that all theoretical bias be eliminated. Confusion over these questions has often led to misdirected attacks; usually the above questions are run together, as in arguments over classification and phylogeny, or in the relationship between classification and evolutionary theory. Not enough attention has been paid to the distinction between the process of classifying and the finished product. Much of this chapter will be concerned with dispelling areas of confusion in this respect. But, before discussing internal issues in any detail, a fundamental difference of approach in taxonomy needs to be characterised.

It is evident that there exists a fundamental dichotomy in taxonomic practice, a dichotomy which is expressed in the attitudes to questions (A), (B) and (C) above. On the one hand there are those approaches that want to maximise theoretical information in classifications and regard classifications as providing explanations. On the other hand, there are those approaches that aim to reject all recourse to explanation and to eliminate all bias. I will call these approaches the THEORETICAL ATTITUDE and DESCRIPTIVE ATTITUDE respectively. These attitudes push up against the limits of the aims and methods of classification, and represent the two possible extremes. In this respect, they represent *ideals* – idealisations or abstractions of various approaches. They document possible extremes of attitude. In practice, however, these two attitudes are never fully realised, although they may be stated as specific goals in taxonomy.

The descriptive and theoretical attitudes are reflected by the taxonomic schools as follows

> *The theoretical attitude* – as exemplified in evolutionary systematics and phylogenetic cladistics.
> *The descriptive attitude* – which is reflected in phenetics and transformed cladistics.

(1)   The theoretical attitude

This argues that at the extreme, classifications are theories, and addresses questions (A1) and (C1) specifically. The aim of all classifications is to carry as much theoretical information as possible, and to reflect the true system in nature.

In the theoretical attitude, it is a prevalent view that

> the ultimate goal of our endeavour is to perform a general integration of functional and evolutionary biology on a causal basis.[4]

The aim is to discover the underlying reality in nature and provide a causal explanation for its existence. This means that a model of the *process* of evolution must be constructed, one which is expressed in the form of a phylogenetic tree (which incorporates a *time scale*). With the subsequent conversion of the tree into a classification, various rules and criteria are adopted which retain the fundamentals of the model of process.

Such an enterprise is not without difficulties, as Agassiz clearly noted

> Nature herself has her own system with regard to which the systems of the authors are only approximations.[5]

Thus, we can only look at what we know about the world, and not at the world itself. And what we purport to know about the world is contained in our best theories. Consequently, there must always be close ties between a classification and the underlying theory. For instance, Hull advocates close links to current theories (especially evolutionary theory), so that all classifications express evolutionary relationships.[6] Given that an underlying theory of process should be utilised in the construction of a classification, it is a short step to arguing that classifications are theories themselves. *A fortiori*, questions over the elimination of bias from a given procedure of classification are prejudged. Theoretical bias should not and cannot be eliminated.

To some extent there had been confusion over questions of storage and retrieval of information and questions concerning the explanatory import of classifications. In specifying that classifications are theories, two senses are possible; either, that the predicates in a classification, e.g. hominoids, convey theoretical import (as opposed to the view in which they are descriptive terms); or, in the conveyancing of information in classifications, where specifying classifications as theories is a shorthand for maximising information content. Both senses are permissable, but ought to be clearly distinguished. Thus, the general assumption[7] that the more information stored in a classification, the more likely it is that an explanation will be found, confuses the two senses. Both information storage and explanation are partially separate issues, since the storage of information does not necessarily presuppose an explanatory component. Quite what the connection is will be examined in Chapters 2 and 3.

Another important aspect of the theoretical attitude concerns the kind of information that can be carried in classifications, e.g. is it evolutionary, typological, or something different? This questions the type of model that is used, and the assumptions made in it. In what follows, I will concentrate primarily on evolutionary information.

The theoretical attitude in taxonomy can be represented schematically:

Model of process of evolution

⇓   results in

Phylogenetic tree

⇓   – what rules do we use?

Classification

Several variations on this theme are found in evolutionary systematics and phylogenetic cladistics, based upon the manner in which the classification is dependent upon the phylogenetic tree. Two positions can be documented: a strict dependence on the phylogenetic tree (Hennig), and a weak dependence, as in Mayr and Simpson.

(i)   *Hennig: classifications as abbreviated statements of phylogeny*[8]

Hennig implicitly argues that classifications are theories, because a cladogram (representing a hypothesis of relationship, i.e. recency of common ancestry) is directly translatable into a classification.

The classification therefore represents an abbreviated statement of phylogeny. As Eldredge & Cracraft put it,

> cladograms are necessary and sufficient for the construction of classifications.[9]

given that

> cladograms are hypotheses about the patterns of nested evolutionary novelties postulated to occur among a group of organisms.[10]

Here, emphasis is placed on classifications as theories in terms of storage and retrieval of information.

(ii)   *Mayr: classifications as theories*[11]

Mayr argues that a classification is a scientific theory, because it has the same properties as all theories in science. Hence

> A given classification is explanatory: it asserts that a group of organisms grouped together consists of descendants of a common ancestor.[12]

A classification must explain why organisms are as they are. A good classification of organisms is automatically explanatory – it elucidates the reasons for the joint attributes of taxa, the gaps separating taxa, and for the hierarchy of categories. In addition, Mayr asserts that classifications should also have a heuristic aspect, which stimulates efforts to recognise homologies and to test the concordance with various types of characters. Finally, classifications must be predictive. Consequently Mayr's emphasis is not only on the storage and retrieval of information, but also on explanation. Classifications should be evolutionary because they

> must necessarily try to reflect the underlying causes of organic diversity.[13]

(iii)   *Simpson: practical classifications*[14]

Unlike Hennig or Mayr, Simpson is not pushing up against the limits of information content and explanatory import in classifications (i.e. questions A and C above); rather, his 'theoretical attitude' is a very much *weaker* version than the *strong* form which I have been documenting up to now. Although some explanatory recourse occurs, this position involves the commonsense suspension of theoretical issues that are regarded as too rigid for practical use. Evidence from the fossil record and various distributional studies is utilised, and there is scepticism over the practical use of explicitly theoretical definitions, such as Mayr's Biological Species Concept,[15] in which the defined unit is one of evolutionary cohesiveness, and the explanation is the mechanism (gene exchange) that produced the cohesiveness. More practical definitions, such as Simpson's minimal monophyly, and his Evolutionary Species Concept[16] are used instead. Any explanatory input is phrased in practical terms, as in Ashlock's definition of the concept of a higher taxon on the basis of gaps.[17]

> A higher taxon is a monophyletic group of species (or single species) separated from each phylogenetically adjacent taxon of the same rank by a gap greater than any found within these groups.[18]

Crucially, Simpson's position does not push up against the limits of the theoretical attitude because of concern over practical issues, such as the classification of fossils.

(2)   The descriptive attitude

This abandons the search for hidden or underlying causes. All aspects concerning the explanation of a system are rejected in favour of a description which does not entail a model of process. Classifications are nothing more than descriptive devices. The descriptive attitude focuses on questions (B) and (C) above, since emphasis is placed on the elimination of theoretical bias from the procedures of classification and the rejection of hidden causes. Methodological considerations are therefore of paramount importance.

Of course, this does not mean that the finished product – the classification – is neutral with respect to theory; all meanings of terms are theory-laden. Rather, it is the methods of classification which attempt to avoid the use of theory. I will return to these points in more detail later. Whether or not all recourse to explanation can be abandoned represents a major area of tension in descriptive attitudes, and analogous to the situation in the theoretical attitude, is partly due to a running together of questions (B) and (C).

The descriptive attitude criticises the theoretical attitude for several reasons: first, it is argued that classifications should be based on the world, and not on theories, because a system of classification based on evolutionary theory would tell us more about that theory, than it would about the world. The use of a model of process presupposes the way we 'carve up' the world. Second, the descriptive attitude condemns the replacement of working hypotheses and theories, i.e. the normal conceptual equipment of the scientific process, by an orthodox matrix of 'theory' (or authoritative explanation) into which the data, as it accumulates, is carefully fitted and arranged.[19] Both of these arguments re-emphasise the point that all theory should be rejected in classification construction. Consequently, classifications should act as *aides-mémoire* or represent simple summaries of knowledge, so that the storage and retrieval of descriptive information can be conveniently executed.

As in the theoretical attitude, the descriptive attitude in taxonomy can be summarised:

Descriptive attitude

implies no model

of change, but a

Structural expectation
(usually a branching diagram)

results in

'Best fit' diagram of nested sets
(a classification)

Since the descriptive attitude confronts the question of the limits in the elimination of bias from our methods, it is equally important to document the justification for the elimination of theoretical assumptions. Indeed, motivation for adopting the descriptive attitude is usually based on a justification of methods and aims. Within the descriptive attitude, two versions can be distinguished – strong and weak.

(i)   *The weak form*

In this form, the use of theory is rejected as far as possible in classification construction. Motivation for the suspension of theory stems from a desire to eliminate question-begging assumptions. The impulse for this attitude stems from worries over circularity that may be present in the theoretical attitude. A classic example cited here is the problem of homology. At its most basic, homology has always been concerned with the comparison of organisms or parts of organisms in the uncovering of relationships. In the literature two basic approaches are present; those that incorporate definitions based on essential similarity and those based on common ancestry. The descriptive attitude has always attacked the latter type of definition on the grounds that the injection of a phylogenetic component in homology[20] results in a circularity,[21] since homology is first defined in terms of phylogeny, and then characters claimed to be homologous are used to infer phylogeny. In cases where theory is involved in the application of various criteria to classification construction, as in this example, the descriptive attitude insists that while science is 'ordered knowledge', it should be empirically orientated with an emphasis on reducing theoretical assumptions in any method. In addition, speculative efforts at generating hypotheses are condemned.

(ii)   *The strong form*

This version goes far beyond the attitudes implicit in the weak form, by aiming to eliminate, wholesale, all explanatory import. Not only are all theoretical presuppositions removed as an end in themselves, but all theoretical interpretations of classifications are rejected. Thus, a theory-neutral method gives a theoretically neutral classification. Classifications are nothing more than simple summaries of empirical data. The early works of the pheneticists (e.g. Gilmour, Sokal & Sneath[22] are closely allied to this strong version. Their rejection of explanatory notions is justified philosophically. The status of transformed cladistics is more difficult to categorise since some of their arguments, e.g. the condemnation of speculative hypotheses, are more in line with a weak version, while other arguments, e.g. the advocacy of process independent patterns, are more akin to the strong form.

The connection between the wholesale elimination of bias and the rejection of explanatory notions will be discussed more fully at the end of this chapter, and the importance of motivation in determining the type of descriptive attitude adopted will be expanded upon by using two examples; Thom's notion of structural stability in developmental biology, and D'Arcy Thompson's mathematical transformations. For the present, the important point in this discussion of weak and strong forms, is that while the former aims to eradicate theoretical bias as far as is practically possible, so reducing question-begging assumptions, the latter is motivated by more philosophical reasons.

Enough has now been said concerning the theoretical and descriptive attitudes in taxonomy to give a general picture. In the remainder of this chapter I want to discuss the implications of each attitude's approach to questions (A), (B) and (C). As well as examining the limits to the two attitudes, I will also clarify possible sources of confusion.

The suspension of theory in the descriptive attitude
The distinction between theoretical and descriptive attitudes highlights an important issue over the question of suspension of theory or change in theoretical standpoint. We are not concerned here with arguments over a change in theoretical standpoint, but rather over the suspension of theory. Seen in this light, it is clear that previous attempts to categorise the transformed cladists as typological[23] (e.g. Charig[24]) are misdirected.

In Charig's typological system there are essentially two steps: firstly, the description of species, and secondly, the construction of a typological classification. These steps may be followed, if desired, by a third, phylogenetic interpretation. Both pheneticists and transformed cladists are supposedly representatives of a typological approach because neither incorporates a direct link to phylogeny. Pheneticists produce a hierarchical classification based on 'overall similarity', while transformed cladists ignore evolution and all other time-related phenomena. They simply arrange groups of organisms into nested pairs on their existing characters, as parsimoniously as possible. The 'natural order' obtained, may, if so desired, be interpreted as reflecting phylogeny. Charig's other system, the clado–evolutionary approach, incorporates the following steps: first, the reconstruction of phylogeny; and second, the construction of a classification consistent with the phylogeny.

Evolutionary systematics and phylogenetic cladistics (including classical Hennigian cladistics) are clearly members of this approach. Now, while Charig's division in methodology has the merit of emphasising the possible relations between classification and phylogeny, it fails to explicate the suspension of theory in a method. The issue of typological and non-typological[25] only concerns the different forms of theoretical assumptions and explanations used and cannot be applied in the manner that Charig intends.

In the first place, 'typological' means different things to different people (e.g. Charig, Hennig,[26] Simpson[27] and Kiriakoff[28]), and all too often, 'typological' as used in present day systematic discourse is a taboo word, usually implying 'unscientific'. Indeed, as Nelson[29] noted, 'typological' is a common perjorative for some 'authorities' who maintain e.g. 'that typological theory or the typological approach should have no part in taxonomy.'[30]

It is all too easy to pick out some deep-seated 'typological prejudice' in one's opponents' system. Yet, the accusation of typology is an odd one if discussing the conditions of evolution (e.g. as in Kiriakoff). Such arguments may just be implying that there is a certain continuity of approach between pre- and post-evolutionary accounts, with the corollary that some typological attitudes are still present.[31]

But there is a more deep-seated flaw in the use of typology for partitioning taxonomic procedure, a flaw that is based on the misuse of historical precedent. In the typology of Owen, the concept of an archetype had an explanatory value because it proved that 'ideas' existed before 'things'. [32]

> The recognition of an ideal Exemplar for the Vertebrated animals proves that the knowledge of such a being as Man must have existed before Man appeared. For the Divine mind which planned the Archetype also foreknew its modifications.[33]

The explanatory force behind the concept of the archetype lay in its conceptualisation as a Platonic essence, or in the case of Agassiz, the Creator's blueprint. Thus

> All organised beings exhibit it in themselves all these categories of structure and of existence upon which a structural system may be found in a manner that, in training it, the human mind is only translating into human thoughts the Divine thoughts expressed in nature in living relatives.[34]

Consequently, both Agassiz's and Owen's typology had a necessary theoretical basis which served to explain the origin of new forms – by Divine intervention for Agassiz, and through 'natural laws or secondary causes' for

Owen.[35] The notion of typological is therefore affiliated to the theoretical attitude, since it does not call into question the rejection of explanation, or removal of theoretical bias. It only concerns the type of theoretical information that is carried in classifications. In contrast, the descriptive and theoretical attitudes call into question the suspension of theory, and the relationship between procedures of classification and scientific theories.

Limits to the storage and retrieval of information in classifications

Over the question of how much theoretical information classifications can carry in practice, two extreme examples have been suggested: classifications as either theories or empirical devices. This concerns the question of whether predicates in a classification carry theoretical import, or are purely descriptive terms. This is, of course, distinct from the problem of maximising information content in a classification, since such information can be either theoretical or empirical. This latter problem concerns primarily the role of classifications as convenient catalogues of all know organisms, while in the case of storing theoretical information, interest is directed at the implementation of theoretical information, and the attainment of the highest degree of theoretical unification.

In what follows, I will argue that both extremes founder on the same difficulty, namely, a misapplied notion of the use of concepts in classification construction. In each case, only one-criterion concepts are used, whereas in reality, classificatory concepts are based on several criteria. Furthermore, the attitudes inherent in talking of classifications as theories or empirical devices are a hangover from positivism. To understand this connection fully, some background over the problems of Universals must be discussed.

The problem of Universals is the problem of justifying the application of the same general term to different individuals or events. Two classic theories of Universals are documented by Hesse,[36] in her discussion of the 'Network Model of Universals'; the *absolute theory* and *resemblance theory*. In the absolute theory, 'P is correctly predicated of an object a in virtue of its objective quality of P-ness.'[37] That is, all the entities which fall under a given term must have some set of properties or features in common, the presence of which makes it correct to subsume an entity under the term. In contrast, the resemblance theory states that 'we predicate P of objects a and b in virtue of a sufficient resemblance between a and b in a certain respect.'[38] The 'certain respect' is the resemblance itself which has been highlighted in the application of P. Thus, a is not used to recongise that it is 'green', but that in recognising a and b, we abstract the resemblance which unites them. In understanding the term 'leaf', we possess a kind of picture of a leaf, as opposed to pictures of particular leaves.

Wittgenstein[39] argued against both of these traditional views in his concept of 'family resemblance' and it is this anti-positivist view that Hesse incorporates into her Network Model. Wittgenstein suggested that all the entities falling under a given term need not have anything in common, but that they are related to each other in many different ways. More precisely, the concept of family resemblance asserts that objects may form a class, to the members of which a

> simple descriptive predicate is correctly ascribed in common language, even though it is not the case for every pair of members that they resemble each other in any respect which is the same for each pair.[40]

Under this view, it is incorrect to look for a common set of necessary and sufficient properties in objects that are classified together, such as 'games'.

The importance of Wittgenstein's 'family resemblance' in biological concepts was first recognised by Beckner.[41] Beckner distinguished between monotypic concepts, in which a class is ordinarily defined by reference to a set of properties which are both necessary and sufficient for membership in the class, and polytypic concepts.[42] In the latter, it is possible to define a group K (of entities) in terms of a set G of properties $f_1, f_2, \ldots, f_n$ in the following way.[43]

Suppose we have an aggregation of individuals (we shall not as yet call them a class) such that:

(1) each one possesses a large (but unspecified) number of the properties in G;

(2) each f in G is possessed by a large number of these individuals;

(3) no f in G is possessed by every individual in the aggregate.

By the terms of (3), no f is necessary for membership in this aggregate, and nothing has been said to either warrant or rule out the possibility that some f in G is sufficient for membership of the aggregate.

Polythetic concepts are therefore 'family resemblance' concepts in that no adequate definition, either by some property of the individuals referred to them, or by any special logical compound of a set of properties, can be found. More importantly, in polythetic classes, it is impossible to assign an individual to K by determining empirically whether or not it is a member of the classes defined by the fs. More information, in the form of the various fs being widely distributed, must be obtained.

The application of these concepts in classification must be clearly specified, since there are several distinct aspects of classification where these concepts can be applied. Firstly, monothetic or polythetic methods can be applied to the problem of partitioning a set of individual organisms into classes, or taxa, which are in some respect homogeneous with respect to specified characters.

Necessarily, this involves an account of the nature of the relation between taxa and individual organisms. It is in this sense that Sneath & Sokal talk of monothetic or polythetic groups (see Chapter 4, p. 109). Secondly, there is the problem of constructing a classificatory system, given as data, the distribution of characters among organisms, in a set of previously established basic classes or taxa. Classifications may be monothetic or polythetic, depending on whether they utilise one or several criteria in their construction. Finally, there is the problem of diagnosis, which involves the placing of individuals within a previously established classificatory system. In examining whether it is possible for classifications to be theories or empirical devices, we will be concerned primarily with the construction of a classification, and not on taxa definitions. It is important to realise that there is *no* connection between an account of the relation between taxa and individual organisms in terms of necessary and sufficient conditions for set-membership, and the logical relations between taxa in a hierarchy.[44] Individuals in a taxa are defined polythetically.

When we talk of classifications as either theories or empirical devices, it is implicitly assumed that only monothetic criteria are used. In the former, since the term 'relationship' is based on phylogenetic descent, then monophyly is used as the definitive criterion for taxa of any rank, since it stipulates that no two distinct branches of a phylogenetic tree should lead into any one taxon. Where classifications are empirical devices, the definition of higher taxa, and their subsequent ranking, must be based on empirical characters, e.g. in defining 'family', 'order' and 'class'. In each case, a term is specified by one criterion alone. But the use of a monothetic method in classification construction is not possible if taxa are polythetically defined, because no one criterion would be privileged. Both the species category and all taxa of other ranks are polythetic (the former with respect to species criteria, and the latter with respect to the set of taxonomic characters), and this reflects the fact that groups of animals possess 'such and such properties', and that other taxa are assigned the places they occupy (i.e. ranking).

In the case of classifications as theories, monophyly is used as a definitive criterion of any rank. But there is a difficulty in defining homology, such that the evidence rarely warrants a definite conclusion about the monophyly of a taxon. Even if there were a simple experimental method for finding out whether or not groups of objects were in some sense monophyletic, monophyly could only be regarded as a diagnostic criterion for taxa in a classi-fication because taxa cannot be defined monothetically. Classifications should therefore be based on other factors, such as the biological description of morphology. It is interesting to note that Simpson was fully aware of these problems, and his minimal monophyly represents an attempt to take into account the polythetic definition of taxa and the polythetic method of classification construction. It should also be emphasised that just because monophyly cannot be regarded as a definitive criterion for taxa in a classifi-cation, this does not preclude the use of a criterion of consistency between phylogeny and classification as a guide in selecting an appropriate classifica-tion.

In strictly empirical classifications, problems occur over the definition of an 'ordinal' or 'class' character. Whether or not a given character is diagnostic for an order or class does not depend upon the character as such, but ulti-mately upon a whole taxonomic system. Consequently, different characters must be involved in classification construction, since all taxa are polythetic in empirical terms.

Clearly, the kinds of constraints on classification systems thrown up by purely logical or formal considerations are not nearly powerful enough to account for the comparatively economical and manageable nature of classifi-cations, or to account for the limited mechanisms of data processing and storage. Classifications are, to a large extent, based on convenience of diag-nosis with the lumping together of organisms with the greatest number of shared properties.

In conclusion, taxa, like many other empirical concepts, cannot be de-fined in terms of any set of properties. It is impossible to specify monothetic criteria for classification construction. Where classifications are theories, no criterion is definitive; it is only diagnostic. As empirical devices, it is impos-sible for any term to function by one criterion alone in a classification; *a fortiori* it cannot be an empirical criterion. With a monothetic criterion, the predicates in a classification would be defined in terms of the classification alone, and these predicates would not necessarily be the same as scientific predicates. This result would be undesirable since the scientific relevance of such a classification would be called into question. (In the theoretical extreme, it is also the case that when the theory changes, then so does the classification.) Empirical classifications would also tend to depend on privi-leged empirical statements, a view that is rejected by the Network Model. These results reinforce the point made earlier that there are no real examples of the extreme viewpoints in practice.

Limits to the elimination of bias

In confronting the limits to the elimination of bias in classification construction, it is important not to confuse this issue with the question of theory-loading, as understood in the philosophy of science. In being opposed methodologically to carrying too much theory, there is no necessary connection with various degrees of theory-loading in the final classification. This distinction is never made explicit in many discussions over description and explanation in classification.[45]

The notion of theory-loading[46] in the philosophy of science concerns the distinction as made in positivism, between observation statements and theoretical statements.[47] Observation terms refer to directly observable entities or attributes of entities, whereas theoretical terms do not. Observation terms include 'blue', 'touches' and 'struck', while terms such as 'electron', 'dream', 'gene' and 'ego' are theoretical. This distinction is emphasised in the positivist notion of theory;[48] theories are 'partially interpreted calculi in which only the observation terms are directly interpreted (the theoretical terms only being partially interpreted).'[49] Against the positivist doctrine of separation of observation and theoretical terms, theory-loading implies that all predicates depend upon theory. The meaning of terms in scientific languages are determined by the theory in which they occur, such that if the same term, e.g. 'mass',[50] 'species',[51] occurs in different theories, it will take on different meanings in the respective theories.

The philosophy of science recognises a whole range of distinctions with respect to theory-loading, and I will briefly discuss Churchland's[52] account to show up the disanalogy between theory-loading in the philosophy of science and the injection of theoretical assumptions in classification construction. Churchland suggests two distinct ways in which all observation vocabularies are prejudiced with respect to theory.

(1)   Intensional bias

A predicate is intensionally biased with respect to some theory when the use of that predicate involves a tacit acceptance of some general assumptions in which that predicate figures. Typically, this may involve the interpretation of data with respect to some theory. Terms such as 'red' are theory-neutral since they are neutral in between the theories that we have, while terms such as 'mass' are not.

(2)   Extensional bias

This is more far reaching than intensional bias in that

> possible observation vocabularies can differ radically in the extensional
> classes into which they divide the observational world.[53]

Thus,

> the range of sensations with which we are endowed does not determine
> a uniquely appropriate vocabulary.[54]

Corresponding description of observable phenomena which are phrased
in two extensionally, closely connected vocabularies may give different
degrees of support to one and the same theory. Consequently, extensional
theory-loading might arise for different types of perceptions, e.g. grue classification,[55] and so reflect incommensurability between different types of
theory-loading. Extensionally biased vocabularies may not recognise the
same animals in a classification of living organisms; what is classified in one
vocabulary may be totally meaningless in another vocabulary. Extensional
bias is directed at the possibility of alternative conceptual schemes. In making
this distinction between intensional and extensional bias, Churchland argues
for the impossibility of the thesis that the meaning of common terms at issue
is given in sensation. The aim is to illustrate the need to drive a wedge
between one's understanding of even the simplest observation predicates,
and one's acquired disposition to make spontaneous singular judgements
containing these predicates in response to whatever range of sensations
nature has given to one.

For our purposes, any analogy between the way that we set up categories
in a classification and in the setting up of theory-laden categories is misdirected, since the former attempts to avoid theory, while the latter concerns
the difference in theory-loading. Theory-loading is related to diagnosis (the
way we apply our categories) and predicate recognition, not predicate construction. Setting up new categories of species is not the same as setting up
new categories of colour. Only the former is possible. This point is important
because normally the distinction between predicates learnt in the context of
a theory and in applying predicates on the basis of appeal to some theory are
run together. But they need not be, and must not be, if, as in taxonomy, the
emphasis is on being opposed methodologically to carrying too much theory.
In advocating the rejection of theory in classification construction, it must be
realised that there still remain implicit theoretical assumptions. *All description
is theory-laden* even if constructed using phenetic criteria.

The conclusion that all description is theory-laden sets up immediate
constraints on the limits to the elimination of bias, since while it may be possible to strive for a theoretically neutral methodology, the conclusion that
classifications are theoretically neutral is not justified. In order to reinforce

this conclusion, I want to concentrate on several examples where, in addition to theory-loading, there are also problems over the elimination of bias in classification construction. The aim will be to examine the possibility of a theoretically neutral methodology. In the elimination of bias we must distinguish between:

(a) the elimination of theory in classification construction and

(b) the elimination of theory involved in application to classification.

On the former question, Sattler[56] has argued that theoretical assumptions can be inadvertently smuggled into the method, such as in the problem of whether or not to use categories in classification construction. This not only involves categories in the sense of the species, genus, family etc, but also in the sense that all theories in comparative morphology (whether typological or phylogenetic) operate with a notion of 'categories'. Generally, this notion involves the discovery of the 'real' (universal) categories which describe fundamental units of form, and the subsequent reduction of the entire diversity of form to these categories. For example, in plant morphology, the natural units include 'phyllome', 'caulome', and 'telome'.[57] In animal morphology the units are 'structural plan' and 'morphotype'.[58] Now, Sattler argues that this idea of universal, morphological categories is a non-starter because of the problem of intermediate forms between typical representatives of the mutually exclusive categories. The use of categories can only account for a restricted number of structures, e.g. 'phyllome'. Further theoretical injection occurs when these categorical theories are combined with the notion of (typological) derivation or (phylogenetic) evolution.

The use of 'categories' in method also arises in the assumption that all observation levels are discrete; thus genetic and morphological levels give information about character states that is relevant only to that level. Or, in the assumption that state changes in characters are either continuous or discrete transformations.[59]

Notwithstanding the problem of theory-loading implicit within the recognition of categories, the issue over categories hits home in the question of data interpretation: any structure which does not fit into the categorical framework of a theory (as in comparative morphology) can be 'explained away' as extremely derived or evolved – it represents the exception which proves the rule. Theoretical input, therefore, occurs in converting observations into data. In addition to this inadequacy in empirical findings, there is also the problem of choosing between contradictory theories due to the underdetermination of theory by the evidence. Competing theories based on categories cannot be judged by the evidence because all the evidence can be found to be consistent with either theory.

In the elimination of theory involved in application to classification, the morphological method provides a good example of further theoretical ineliminability in method. In his discussion of idealistic morphology, Zangerl[60] distinguishes between two approaches in the elucidation of morphological concepts. There are those morphological concepts which are broadened and transferred from the methodological to the theoretical level, as in the injection of a phylogenetic component, and there are those concepts which are further developed within the realm of their applicability, i.e. within comparative anatomy. In the former the results are of a theoretical nature, while in the latter, they supposedly remain strictly factual.[61] Zangerl concentrates his discussion within the realm of comparative anatomy (Fig. 1).

Zangerl believes that his morphological method is an empirical science which aims to uncover the conformity to various kinds of basic patterns in nature. Thus, morphology should deal with the investigation of the normal phenotypically expressed structural design of organisms, through the comparison of form (whether between adult recent and adult recent, or adult recent with fossil recent, or adult recent with embryonic). All morphological

Fig. 1 Zangerl's conception of experimental and non-experimental sciences.[62]

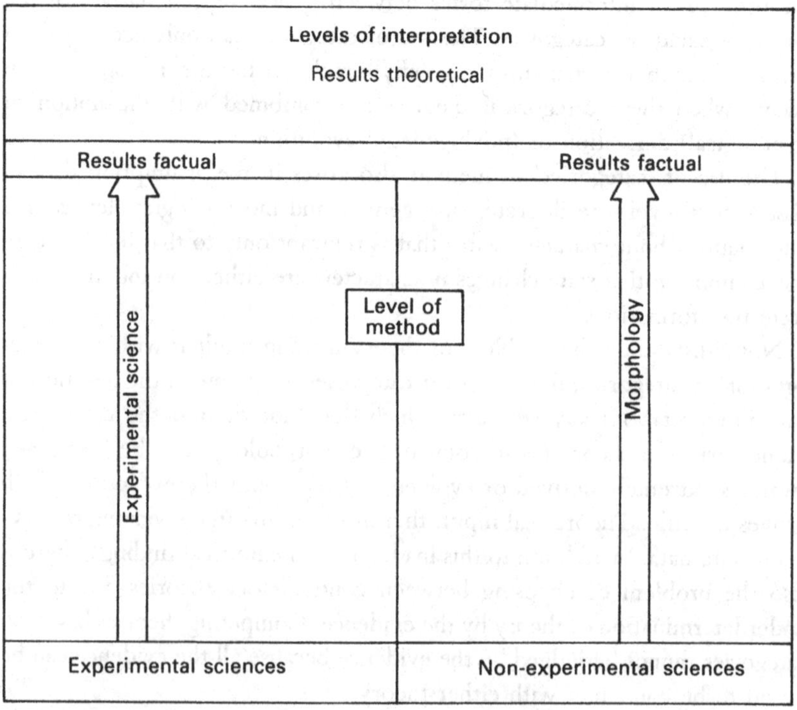

concepts purport to express observed relations. These relations are facts, verifiable by any subsequent observer. Hence, morphological concepts are factual generalisations derived from observed structural relations and they do not and cannot carry phylogenetic implications concerning common ancestry: these are restricted to the realm of the theoretical.

Unfortunately, Zangerl's distinction between the theoretical and methodological levels does not purge all theory from the latter. Theoretical assumptions are implicit in his concepts of the structural plan and morphotype. Theoretical ineliminability in the structural plan occurs in a comparison of the parts of organisms topographically:

> The conformity of a design in the topographic (spatial) relationships of the parts of an organism to the body as a whole is called the structural plan,[63]

while the morphotype is a theoretical concept:

> the norm is the morphotype which is an abstraction of the actual form variety within a group of organisms of the same structural plan.[64]

Structural plans are said to express relations of far greater scope than do morphotypes. Thus the morphotype of the skull of the classes Mammalia, Aves, Reptilia and Amphibia are all different, but the structural plans expressed in these four morphotypes are essentially the same.

The use of such concepts in the morphological method places great emphasis on the uncovering of the *gesetzmassigkeiten* (conformity to various kinds of basic patterns in nature) as expressed in various empiric principles, such as Dollo's principle of irreversibility in evolution. Zangerl believes these empiric principles to be strictly factual because they express various relationships whose reality lies beyond the activity of the human mind, and so cannot be rejected. In the face of counterinstances, their range of applicability may be decreased.

Yet, is it fair to claim that an empiric relationship is a fact in the face of contradictory evidence? I think not, because such empiric principles are really rules of thumb. Evolutionary assumptions can easily be smuggled in with rules of thumb, so that Zangerl's claim of facticity is greatly imbued with theory. This problem of interpretation at the level of method is not adequately recognised by Zangerl,[65] and this constitutes a serious problem for the descriptive attitude, which requires a total elimination of bias in method. Further theorising is necessary in the morphological method in decisions of comparison: when is it permissable to compare, say, parts of organisms? Such problems undermine the desire for a theoretically neutral method, and are confronted in the problem of homology, to which we shall now turn.

The problem of homology

The problems arising in homology bring into greater relief the links between the descriptive and theoretical attitudes, since homology is one area where they meet. A principle area of tension between the two attitudes concerns decision making at the same level of analysis. Are they using the same criteria for recognition of homology? My discussion will focus on this tension.

In the theoretical attitude, definitions of homology are couched in terms of common ancestry. Evolutionary systematists define homology as 'structural similarity due to common ancestry'.[66] Worries over circularity, where similarity is used to infer common ancestry, which is then used to validate the similarity in homology, prompted Bock to reject the term 'similarity' in his definition:

> A feature (or condition of a feature) in one organism is homologous to a feature (or condition of a feature) in another organism if the two features (or conditions) can be traced phylogenetically to the same features or conditions in the immediate common ancestor of both organisms.[67]

These definitions aim to distinguish homologous similarity from convergent similarity. In contrast, phylogenetic cladists are not concerned with this distinction, because they use a model of evolution that is divergent. For instance, in Hennig's definition

> Different characters that are to be regarded as transformation stages of the same original character are generally called homologous.[68]

Later phylogenetic cladists attempted to define homology in terms of monophyly,[69] but failed because monophyly is to do with groups, not characters. More recently, Eldredge & Cracraft have conceptualised homology as synapomorphy:

> homologous similarities are inferred inherited similarities that define subsets of organisms at some hierarchical level within a universal set of organisms.[70]

In the descriptive attitude, the aim is to eliminate the phylogenetic component, with special emphasis on similarity. In phenetic definitions further emphasis is on practical applicability:

> The relation between parts which occupy corresponding relative position in comparable stages of the life histories of the two organisms.[71]

But early phenetic definitions which argued against a phylogenetic concept of homology on the grounds that it was 'not susceptible to direct proof but only to proof-by-inference',[72] had to be relaxed because it was realised that theory and inference were also necessary in phenetic procedures.[73] Later phenetic definitions were less precise:

> Homology may be loosely described as compositional and structural correspondence.[74]

Transformed cladists also use a notion of similarity in homology, 'special' similarity, as characterised in their concept of synapomorphy. Thus,

> homology can . . . . . be defined as the relation characterising (natural) groups.[75]

There are several important differences in the characterisation of homology in the descriptive and theoretical attitudes.[76] In the theoretical attitude, there is an implicit model of process, conveyed by the term 'common ancestry' in evolutionary systematics, and by *a posteriori* synapomorphy in phylogenetic cladistics. But as Zangerl pointed out, the concept of homology can no longer function as a tool because of the input of causality. We do not and cannot *a priori* know anything about the causality of a given structural relationship between different parts of different organisms. Homology and the cause of homology are run together in the theoretical attitude. Contrastingly, the descriptive attitude is based on a structural approach – the recognition of similarity.[77] Over and above this problem of specifying homology, there is the question of the role that homology has in each attitude, and its methodological status. These two problems are closely interconnected.

Van Valen[78] has recently argued for homology as similarity by virtue of common ancestry. But how do we know that we have common ancestry in the first place? The simple answer is that we don't, since homologies are supposedly worked out first, and then linked to a phylogenetic tree. What we really have is a presupposition of a function of change with time,[79] or more specifically, a model of process based on pattern estimation, aimed at specifying the transformation of one set of characters into another set. Prior to any assessment of common ancestry, there must be some decision over pattern matching. Any view of change through time requires a prior assessment of similarity. Jardine[80] emphasised this point about pattern recognition in his distinction between topographic homology (corresponding in relative position) and phylogenetic homology, the latter being a special case of the former.

Given that pattern recognition is procedurally prior to an assessment of common ancestry, the view (in the theoretical attitude) that homologies are first calculated, and then linked to a phylogenetic tree is clearly spurious. More realistically, homology is *a posteriori* with respect to phylogenetic tree construction, in that there must first be a tree, which is then used to specify the ancestry of homologues. But any separation of procedures in this manner is misleading since the problem of homology and tree construction is really one and the same problem. One model (based on change in time) leads to one problem: elucidating homologies and phylogenetic trees.

The emphasis on procedure is different in the descriptive attitude. One problem arising out of the theoretical attitude concerns subjective probability decisions over characters. These subjective decisions are usually based on biological rules, as specified in the model of process. In the following a descriptive attitude, these decisions are regarded as misleading because there must be some criterion for the choice and comparison of characters prior to the superimposition of a model of process. In the descriptive attitude, emphasis is rightly placed on the problem of when to make comparisons, since homology is first and foremost a problem of assessing the probability of similarity. But several problems still remain in the descriptive attitude; firstly, how do we compare two things to establish the resemblance between them? Secondly, what degree of resemblance is considered sufficient to justify the relationship of homology? In the first case, if correspondence in relative position is taken as a basic criterion for parts to be homologous, there still remains some uncertainty over when to make comparisons. For instance, in defining homology as

> The relation between parts of organisms which are *regarded* as the same,[81]

there is a hidden point: what is the status of *regarded*? Clearly, there is a presupposition of a model of decision making for when comparisons should be made. Thus, there is always an implicit assumption that characters are comparable. In the theoretical attitude this problem is predetermined by the decision that only those characters which give evolutionary information are of any interest. In the descriptive attitude, there are always theoretical presuppositions creeping in, because some decision for comparison has to be made.

This practical ineliminability of theoretical assumptions is clearly shown in Jardine's[82] discussion of skull roof matching in Rhipidistian fishes. Jardine concluded that the matching of parts always incorporated a fundamental premise, e.g. the Swedish school of paleontology, typically represented by Jarvik,[83] *a priori* assumed a decrease or increase in the number of bones, which was consistent with a view of evolution being conservative.[84] Jardine concludes:

> that the hypothesis of at least partial conservation of skull topography

in phylogeny is a *sine qua non* for the use of homology as a basis for inference about the morphology of the skulls of ancestral organisms.[85] Although the example cited here for structural correspondence (based on where parts have boundaries) resorts to an examination of transformations through time, any attempt to define homology solely in terms of resemblances is misleading, because in practice additional criteria, such as consistency with the developmental relations between parts, or similarity in composition, must always come into play.[86] These additional criteria may also have evolutionary significance. Correspondence in relative position is therefore only a necessary, not sufficient, criterion for determining homologies based on similarity. To argue for compositional correspondence between features, as Sneath & Sokal do, automatically brings in a theoretical assumption, since for homologous parts to be of similar substance implies continuity of change.

So far we have examined the problem of practical ineliminability of theoretical assumptions in the descriptive attitude.[87] Clearly, there are very real limits to the elimination of bias. But there still remains the problem of the degree of resemblance required to justify the relationship of homology. Should it be absolute identity or quantitative resemblance? In the former case, Inglis[88] argues for the either/or status of homologues, such that organ X is either homologous to organ A or organ B. As we have seen, such quantitative definitions are incapable of dealing with continuous variation.[89] Sattler[90] has therefore suggested a semi-quantitative definition based on the question of whether organ X is more homologous to organ A or organ B. But in adjudging whether organ A is more homologous to organ X, there is still an element of theoretical ineliminability, since there has to be a decision of how similar organ A should be to organ X.

Within the problem of homology, it is evident that there is a dichotomy in attitudes, such that different procedures for decision-making are made at different levels. On the one hand, homology is regarded as being essential to tree construction, with an underlying model of process. This model is always chronistic, i.e. based on time, and utilises appropriate biological rules. On the other hand, homology is based on a structural model for transformation, and rules for the transformation of one form to another are specified (as in D'Arcy Thompson's transformations). This dichotomy between an emphasis on biological rules as opposed to structural models may give very different ideas over homology. Both have drawbacks. In the theoretical attitude, procedural priority is not given to pattern matching when it should be, while in the descriptive attitude, problems over the elimination of bias do not justify the desire for a theoretically neutral methodology.

Limits to the rejection of explanation

The strong version of the descriptive attitude specifies the rejection of explanation. In Darwinian terms this means that the evolutionary explanation must be decoupled from evidence based on the hierarchic system in nature. Is this a viable proposition? In examining this question and highlighting the limits to the descriptive attitude, I want to discuss several examples from outside of taxonomy, which will clarify some of the issues involved.

(a)   *Thompsonian transformations*

D'Arcy Thompson's transformations represent a full-blooded attempt at abandoning causal networks in favour of a mathematical-descriptive approach.

> My sole purpose is to correlate with mathematical statement and physical law certain of the simpler outward phenomena of organic growth and structure of form, while all the while regarding the fabric of the organism, *ex hypothesi*, as a material and mechanical configuration.[91]

Central to Thompson's approach is the construction of transformation grids (Fig. 2 (a), (b)) or co-ordinate diagrams which are devised for the comparison of related forms or of successive growth stages. One figure is overlaid by a rectangular grid, and in another figure to be compared with it, the grid lines are distorted so that they pass through the corresponding morphological points.

Thompson found that for many related creatures all the apparently distinct, external differences between them could often be accounted for by a single integral transformation of co-ordinates, such as an elongation or a rectangular expansion of one area. For example, the human skull (when superimposed upon a single two-dimensional grid) could be easily transformed to resemble an ape-like skull by changing a single aspect of one co-ordinate (Fig. 2 (b)). Other examples included Acanthopterygian fish and crocodile skulls.

Thompson's method is descriptive in that the transformations represent an expression of relations;[94] the emphasis is on the configuration of related forms rather than in the precise definition of each, or even in the source of beginnings and ends of actual processes. Not only are the morphological correspondences *observed*, but the models are also typological:

> the form of related organisms . . . show(s) that the difference between them are as a general rule simple and symmetrical, and first such as might have been brought about by a slight and simple change in the system of forces to which the living and growing organism was exposed.[95]

Thompsonian transformations attempt to reject the use of theoretical pre-suppositions in the method, and the motive for this is based, in part, on the naturalistic tradition – if we look at nature, then we can see new things. Such regularities as may be perceived in nature cry out for an explanation. This explanation can be based on evolutionary theory; thus, in converting the morphogenetic patterns to phylogenetic patterns, an evolutionary postulate has to be introduced.

> If two lives exhibit the same Bauplan, then there is an ancestor common to both which exhibits a Bauplan which is exhibited by all three lives.[96]

In this context, the

> . . . logical role of the theory of evolution is to explain the results of morphological research by introducing certain fundamental beliefs or postulates which have these results as their logical consequences.[97]

However, with an emphasis on recurrent patterns, the point being made by Thompson is that biologists readily assume there to be a proper expla-nation, [98] and that it is invariably evolutionary. (An analogous example is Gould's[99] discussion of struts in bones.) But, instead of searching for explan-ations, and incorporating related assumptions into the method, Thompson argues for a method that gives a 'structural fit' or correspondence, which can be decoupled from an explanatory component. This does not mean to say that the regularities exposed in nature are non-explicable, only that their dis-covery does not necessarily presuppose an explanatory element.[100]

Over and above this initial problem, there are further limitations in the mathematical–descriptive method. Sneath[101] has argued that the transforma-tion grids are of limited use because it is difficult to draw the grids with the required accuracy, and to express the deformation in mathematical terms that would allow general conclusions to be drawn. Sneath has also criticised Thompson for underestimating how much can be done with a few distor-tions that are based on only, say, three to four characters. As a pheneticist, Sneath argues that the Thompsonian approach fails to tackle properly the question of whether equal weight should be given to characters, and whether or not the characters are independent of one another. If the trans-formations are based on equally weighted characters, then the major trends in the grids are an expression of overall difference. If, however, the transfor-mations allow for redundancy of characters then the trend is an expression of the major difference.[102] In the latter case it may be necessary to increase the estimation of the number of mathematical transformations required, with the result that some transformations will be very complex.

Bracketing out the phenetic attitudes implicit in Sneath's criticisms, it is evident that without theoretical constraints, it is possible to follow through any type of deformation; there are no limits on what can be changed or transformed. Consequently, recourse to some kind of explanation *has* to be made, if only to define the boundaries within which a deformation is said to have occurred.

(a)

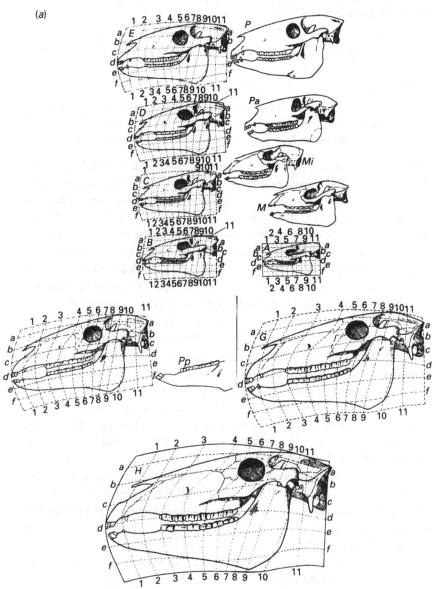

(b)   Structural stability and morphogenesis

Thom[103] advocates a descriptive method in his mathematical models of the developmental process. Description occurs at the *local*[104] level since it is confined to the observables at hand.[105] Thom's models are topological,[106] qualitative rather than quantitive, and do not depend on any scheme of underlying casual explanation. As Thom explains:

Fig. 2 Transformation grids.[92] In (*a*) there is a comparison between A, a skull of *Hyracotherium* (Eocene) and H, the modern horse; the latter is represented as a co-ordinate transformation of that of *Hyracotherium*, and to the same order of magnitude. B – G are imaginary, intermediate reconstructions between stages A and H. Skull C is compared with M, *Mesohippus* (Oligocene); E with P, the skull of *Protohippus* (Miocene); Pp, lower jaw of *Protohippus plaudium*, with F. Both Mi, *Miohippus* and Pa, *Parahippus* show less perfect agreement with C and D. In (*b*) four hominoid skulls are shown in saggital section, with transformation grids. All refer to A, *Homo sapiens*.[93]

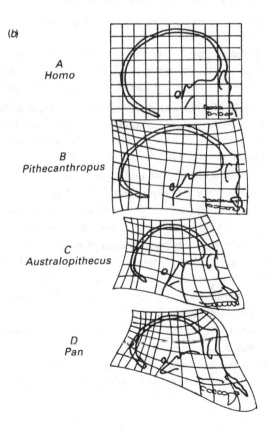

(b)

A
Homo

B
Pithecanthropus

C
Australopithecus

D
Pan

> One essential feature of our use of local models is that it implies noth-
> ing about the "ultimate nature of reality", even if this is ever revealed
> by analysis complicated beyond description. Only a part of its manifes-
> tation, the so-called observables are finally relevant to the macroscopic
> description of the system. The phase space of our dynamical model is
> defined using only these observables and without reference to any
> more or less chaotic underlying structures . . . . . But we cannot hope
> *a priori*, to integrate all these local models into a global system. If it
> were possible to make such a synthesis, man could justifiably say that
> he knew the ultimate nature of reality, for there could exist no better
> global model.[107]

Thom's aim is to produce an abstract, purely geometrical theory of
morphogenesis which is *'independent of the substrate of form and the nature of
forces that create them'*.[108] No reference is made to underlying structures. This
approach is clearly mathematical–descriptive, geared to abandoning the
underlying causal network. Furthermore, this geometrical analysis of the
problem of morphogenesis is based on topological analysis,

> for topology is precisely the mathematical discipline dealing with the passage from
> the local to the global.[109]

The motivation for Thom's approach comes from his radical mistrust of
reduction. He totally opposes the atomistic ideal of constructing one theory
based on the combination of elementary particles and their interaction.

> We must reject this primitive and almost cannibalistic delusion about
> knowledge, that an understanding of something requires first that we
> dismantle it, like a child who pulls a watch to pieces and spreads out
> the wheels in order to understand the mechanism.[110]

In this sense, Thom rejects the use of underlying causal notions; in its
place there is an alternative form of 'explanation', based on an attempt to
attribute a formal geometrical structure to a living being, and to explain its
stability. This is not a causal explanation, but rather a mathematical description divorced from any underlying theory, based on practical considerations.

Within this characterisation of explanation, there is an underlying tension: while his approach is superficially mathematical–descriptive, it is not
strictly a strong form of the descriptive attitude. There still remains, in the
final analysis, a notion of explanation within his description of
morphogenesis – a geometrical explanation. This may not necessarily be a
scientific explanation, but it nonetheless has explanatory import.

In conclusion to the question of whether it is possible to reject an explanatory component, it is evident that the ideal of a pure descriptive attitude

founders on the problem of being implicitly theoretical, thereby requiring a recourse to explanation. Without any theoretical input it would be impossible to specify the boundaries for description, or even the recognition of the components in the first place.

Conclusions

In distinguishing between the theoretical and descriptive attitudes, I have documented contrasting views; either that theories are essential for the construction of a classification, or whether a taxonomic system is desirable in building a theory. In each addition, two versions were suggested: in the theoretical attitude a strong form which assumed a direct translation between phylogeny and classification, and a weaker form in which this relation was not one-to-one. In the descriptive attitude, the strong form rejected fundamental assumptions and postulates believed to be incorrect; the underlying theory is totally 'at sea'. The weak form, however, did not call into question the deep-lying assumptions but argued that because of the sensitivity of techniques to the underlying theory, it was impossible to construct a method that did justice to the causal input. Theoretical assumptions were rejected on the grounds that they begged the question.

In examining the limits inherent within these two attitudes, we have seen that:

(i) classifications are neither theories nor empirical devices;

(ii) over and above the problem of theory-loading, there are real limits to the elimination of bias in classification construction which call into question the viability of the strong form in the descriptive attitude;

(iii) it is impossible to eliminate explanation because it results in either the impossible suspension of scientific beliefs, which in turn lead to internal contradiction, or the difficulty of specifying the limits of such an approach without recourse to some form of explanation.

In *practice*, then, the limits to the two attitudes are not pressed up against. In the descriptive attitude the problem of the practical eliminability of underlying theoretical assumptions cannot be adequately resolved, while in the theoretical attitude, it is not true that the introduction of more and more theoretical information leads to an explanation.

We are now in a position to see whether the various taxonomic schools are coherent in formulating their aims and methods. In the next five chapters I will be concerned with examining a middle ground: that is, given the limits that are present, are the aims and methods of taxonomy sensible? Two types of inconsistency will be shown up in this discussion. Firstly, the aims and purposes are not concordant with the methods. Both Hennig and Mayr are guilty parties here. Secondly, both the transformed cladists and early pheneticists do not practice what they preach. In this respect, it is possible to judge each taxonomic school on the basis of internal consistency.

PART II

The status of theoretical classifications

2

Evolutionary systematics and theoretical information

Introduction

The main issue to be confronted in this and the following chapter is how much theoretical information classifications can carry in practice. As we have seen, to talk of classifications as either theories or descriptions alone can be construed as a shorthand for the conveyancing of information in classifications, and that this sense must not be confused with an alternative, in which the predicates in a classification convey theoretical import or are purely descriptive terms. Both evolutionary systematists and phylogenetic cladists make a point of arguing that their respective classifications store the greatest amount of theoretical information. This claim will be examined in detail with a view to clarifying the prejudice that the more information that is fed into a classification, the more likely a correct explanation will be uncovered. The reasons for the failure of this prejudice will be discussed and it will be argued that the aims as set out, especially in Mayr and Hennig, are not concordant with the methods. Indeed, some of the aims do not appear to be very sensible.

My discussion will be structured in terms of giving an account of:

(1) the purposes of biological classification;
(2) the manner in which each of the above systems (i.e. Mayr, Simpson in this chapter and Hennig in Chapter 3), attempt to fulfill that purpose;
(3) and whether or not these methods accomplish the purpose for which they are intended.

Before examining these issues, a short account of evolutionary systematics is necessary.

Evolutionary systematics represents the mainstream position of most practising taxonomists, and its scope is defined by the classical works on the subject – Simpson's *Principles of Animal Taxonomy*, and Mayr's *Principles of Systematic Zoology*. The roots of evolutionary systematics are to be found in Huxley's *The New Systematics* and Simpson's *Tempo and Mode in Evolution*. This diverse and partially intuitive system of classification is mirrored in its varied nomenclature: at any one time labels such as 'traditional', 'conventional', 'gradistic', 'synthetic', 'eclectic' and 'syncretistic'[1] have been applied.

The aim of evolutionary systematics is to discover groups of organisms in nature, groups that are the product of evolution. Hence, biological classification should reflect the *one* theory in biology – namely evolution, in all its aspects. As Bock puts it:

> . . . the underlying theory for biological classification is that of organic evolution and in particular that of the synthetic theory developed over the past forty years. No theory of biological classification exists in itself; nor can classification be developed in the absence of theory.[2]

The methods employed in uncovering such aims are inherently arbitrary. All evolutionary systematists recognise that intuition and subjective judgement enter into the process of classification. Indeed, Simpson sees the process of taxonomy as akin to art, based on 'human contrivance and ingenuity.'[3]

Evolutionary systematists regard classification as a multi-stage procedure,[4] which can best be viewed as follows.

(1) Organisms are selected for classification.

(2) Data are assembled about the organisms.

(3) The organisms are sorted into taxonomic units, demes (local populations), subspecies or species as may be appropriate. Data on polymorphic forms and all kinds of variation are analysed.

(4) The characteristics of the varying units are compared, with special attention being paid to the kinds and degrees of resemblances.

(5) The relationships revealed by comparison are interpreted in terms of basic taxonomic concepts such as homology, parallelism, primitiveness and specialisation.

(6) From (5) inferences as to the evolutionary patterns amongst the study populations are made. A phylogenetic tree is drawn.

(7) Conclusions on affinities, divergence etc., are translated into hierarchic terms, the various groups of organisms being assembled and divided into taxa of various ranks.

The principal methodological rule that evolutionary systematists employ is that there must be a systematic relation between a classification and a phylogeny. This is possible because both classifications and phylogenetic reconstructions are derived from the same basic evidence: a comparison of more closely or more distantly related species and the evaluation of similarities (or differences) in individual characters, i.e. comparative character analysis. Thus, phylogenies and classifications are based on a study of natural groups found in nature, groups having character combinations that one would expect in the descendants of a common ancestor. The systematic relation between the formal classification and the phylogenetic diagram is not one-to-one, because each contains information not possessed by the other. Phylogeny is not based on classification and nor is classification based on phylogeny. Instead, the relationship between phylogeny and classification is more subtle and it is in understanding this relation that the real impulse of evolutionary systematics can be grasped.

In studying the phylogeny of a group of organisms, three factors are of interest:

(a) the pattern of phylogenetic branching;
(b) the amount and nature of evolutionary change between branching points;
(c) the distribution of characters in populations of ancestral organisms.

In any classification, the evolutionary systematist attempts to maximise simultaneously the information expressed in factors (a) and (b); the branch points and the evolutionary history of a particular branch (e.g. whether or not it has entered a new adaptive zone and to what degree it has expressed a major radiation). An important corollary of this is that the phylogeny represents a model of evolutionary change that presupposes *gradualism*. Significant evolutionary change happens through the *in toto* transformation of lineages. Clearly, evolutionary systematists are not neutral on the mode of speciation: speciation cannot occur without phyletic evolution.[5] This is an important point when it comes to representing such information in a classification, since it is presupposed by the question of how much theoretical information classifications can carry.

It is over the third factor, however, that a fundamental split in method occurs within evolutionary systematics. Two methods of representing evolutionary information in classifications have been documented.

Phylogenetic weighting[6]

The fist position, as proposed by the neontologist Ernst Mayr,[7] follows Cain,[8] and Cain & Harrison[9] in adopting the notion of phylogenetic weighting. Cain originally argued that rare characters should receive more weight because they are better indicators of evolutionary relationship between taxa, or groups of organisms. Mayr followed this up by arguing that the correct method of weighting characters was phyletically;

> Weighting . . . can be defined as a method for determining the
> phyletic information content of a character.[10]

For Mayr, different characters contain very different amounts of information concerning the ancestry of their bearers. This is meant to be a strictly
empirical procedure, in that evidence derived from character distributions in
populations of ancestral organisms is used in determining not only the pattern of branching points, but also the amount of evolutionary change along a
single lineage (i.e. between the branch points).

Phylogenetic modification[11]

The second method, as set out by Simpson in *Principles of Systematic
Zoology*, does not emphasise the information conveyed in the weights
accorded to ancestral character distributions, but advocates information storage via the modification of classification to phylogeny. An evolutionary classification is obtained via the modification of evolutionary branching
sequences so as to render it consistent with evolutionary divergence and
rates of evolution. In this process of phylogenetic modification, it is assumed
that fossils play a central role in the procedure of dividing various groups of
organisms into taxa of various ranks. Crucially for Simpson, a study of fossils
will establish evolutionary branching sequences, and the resulting consistency between phylogeny and classification is used as an expression for relating fossil organisms with present day organisms. This is an important point
because Simpson believes that fossils play an essential role in determining
evolutionary relationships, and that the information derived therefrom must
be stored in a classification. As we shall see later, this is a point that is
attacked by all cladists.

(a)   *Phylogenetic weighting.*

For Mayr, the precise meaning of the term 'relationship' is all
important in the process of phylogenetic weighting. Relationship is based on
inferred genetic similarity as determined by both distance from branching
points and subsequent rate of divergence. The justification for defining relationship as the inferred amount of genotype appears straightforward: for
what is of primary interest is a taxon which expresses its evolutionary role,
its systems of adaptations, and all the correlations in its structures and characters. All of these are ultimately encoded in the genotype. By concentrating
on inferred genetic similarity, Mayr assumes that fundamental difficulties
over the recognition of parallelism, convergence and mosaic evolution will
be solved at the level of the gene. In contrast to Simpson, Mayr implicitly
rules out the use of fossils in classification construction because of his
emphasis on the genotype.

The classic example cited in Mayr's concept of relationship is based on a comparison of birds, crocodiles and lizards. In Fig. 3,[12] the inferred genotypical distance from A is 15% for B, 10% for C, and 70% for D. Between the genomes of B and C there is a maximal genetic difference of 25%, while for genomes C and D there is a difference of 60–70%. On the evidence of inferred genetic similarity C is much more closely related to B than to D, *despite* the fact C has a more recent common ancestor with D, than with B. Substituting B, C and D for lizards, crocodiles and birds respectively, and given that crocodiles have a more recent common ancestor with birds than with lizards, then crocodiles belong to the same evolutionary clade as birds. Crocodiles and birds constitute a group of common genetic origin in that their lineage branches are a direct result of lineage splitting. Clades are therefore monophyletic groups (of any magnitude) in which all members are descended from a common ancestor. Clades are normally contrasted (Fig. 4) with grades[15], because the latter are a product of anagenesis – evolutionary advance along a lineage, while the former are produced through cladogenesis, the splitting of branches or lineages. Anagenetic episodes usually create organisms with novel characters and abilities beyond those of their ancestors. Organisms characterised by such new functional abilities are said to have achieved a new grade. The development of endothermy, for example, created a new grade. Grades can be defined simply as units of anagenetic advance, which are either monophyletic or polyphyletic (groups derived from two or more ancestral lineages) as in endothermy, since both birds and mammals have achieved this condition independently. Returning to Fig. 3, we can see that crocodiles do not belong to the avian grade because they are not a group similar to birds in general levels of organisation. Instead, crocodiles are an example of the reptilian grade.

Fig. 3 Mayr's concept of relationship.[13]

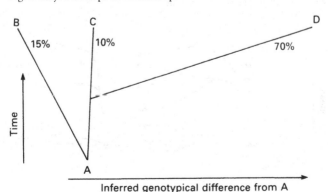

Inferred genotypical difference from A

This distinction between clades and grades is of great importance because it highlights the processes of evolution that all
evolutionary systematists believe to be taking place. Both clades, as expressions of branching sequences in evolution, and grades, representing levels of
adaptively unified complexes of characters, must be incorporated in classifications. Mayr and Simpson are in total agreement on this point: classifications must reflect all aspects of evolution – both branching sequence and
evolutionary divergence. It is in the types of evidence used to construct evolutionary relationships that their differences lie. This alone has significant
implications for the representation of such information in classifications, and
consequently in procedures of ranking.

For Mayr, classification construction requires two steps:

(1) the grouping of lower taxa (usually species) into higher taxa;
(2) the assignment of these taxa to the proper categories in the
taxonomic hierarchy (ranking).

Referring back to the bird, crocodile and lizard example, crocodile and lizards are grouped together into one taxon, while birds constitute the other
taxon. Because birds represent a higher grade than reptiles, they are assigned
to a higher rank than if they had been defined purely cladistically. The classification is as follows:

> *Taxon 1*   (C + B   REPTILES)
>
> (a)   Taxon C
> (b)   Taxon B

Fig. 4 Relationship between anagenesis (evolutionary advance or change) and
cladogenesis (branching of a lineage); between clades (branches) and grades
(levels of functional and morphological complexity). Time runs to the right and
evolutionary change runs up or down. The branching pattern represents the
evolution and slitting of lineages, some of which undergo periods of rapid
anagenesis. Grade 2 has been achieved twice by independent lineages, each of
which has then split to form a clade.[14]

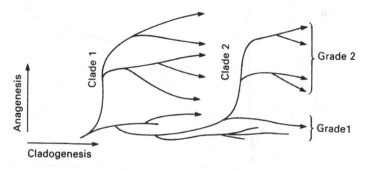

Taxon 2   (D   BIRDS)

Mention must be made of the criteria that Mayr utilises for the delimitation and ranking of taxa. Such criteria are based on the products of the evolutionary process, and include distinctness between taxa, the evolutionary role (uniqueness of adaptive zone), the degree of difference – 'distance' between the means of the two groups of species, the size of the taxon, and finally the equivalence of ranking in related taxa.

(b)   *Phylogenetic modification.*

Simpson's method relies on the modification of results by phenetic methods so as to ensure *consistency* with what can be inferred about the evolutionary branching sequence of the populations studied. This means that a taxonomic hierarchy for a group of organisms is said to be consistent with a phylogeny if all its taxa of supraspecific ranks are *minimally monophyletic*; that is, the

> derivation of a taxon through one or more lineages (temporal successions of ancestral–descendant populations) from one immediately ancestral taxon of the same or lower rank.[16]

The level of monophyly is specified by the category of the lowest ranking, single taxon immediately ancestral to the taxon in question. Thus, a taxon is said to be monophyletic at rank *J* if all organisms referred to it are directly descended from organisms referred to not more than one taxon of rank *J*. The importance of minimal monophyly as a criterion of consistency is clear, given the fact that any stricter definition of monophyly, such as in Hennig's work, requires that *all* species derived from the stem species be included, as well as the stem species itself (see p. 69). Under strict monophyly, fossil species cannot be conveniently placed in a classification that requires descent from a group of species back to a common stem species. Furthermore, because it is possible to 'shuttle' from minimal monophyly to strict monophyly,[17] consistency between classification and phylogeny can have a weak or strong formulation. The weak form, based on minimal monophyly, caters for the convenient inclusion of fossil species, while the strong form, based on strict monophyly, does not. Depending on how fossils are treated, there is a continuum between Simpsonian and Hennigian methods of classification construction.

Several provisos must be stipulated in Simpson's notion of consistency. In the first place, the relationship of consistency is not one-to-one: several phylogenetic branching sequences may be consistent with one taxonomic hierarchy and several taxonomic hierarchies may be consistent with one phylogeny. Secondly, it relates phylogeny to classification. Simpson is at times unclear in detailing this; as far as can be made out, Simpson's view is to set up a classification first, then check it against a phylogeny, and amend if necessary. In this respect, Simpson talks of phylogeny and classification as being separate, rather than closely connected. This is a problem which the phylogenetic cladists confront in detail. Finally, the feature of phylogeny which interests us, if we are to ensure the consistency of the classification of groups or oganisms with their phylogeny, is the tree-form and not the absolute or relative times of branching.

From what has been said so far, it is evident that Simpson's classificatory procedures are pragmatic, geared to the inclusion of fossil studies. Nowhere is this practical impulse clearer than in Simpson's criteria of adequacy for classification. Three criteria are relevant.

(1) Criteria related to objectivity, reality and non-arbitrariness of the fundamental units of classification. Simpson rejects the terms 'objective' and 'real' because they are misleading, and uses the term 'arbitrary' instead.

> I propose to call the taxonomic procedure arbitrary when organisms are placed in separate groups although the information about them indicates essential continuity . . . or when they are placed in a single group although essential discontinuity is indicated. Conversely, procedure is non-arbitrary when organisms are grouped together on the basis of pertinent essential continuity and separated on the basis of pertinent essential discontinuity.[18]

Species are the fundamental units of classification, and under Simpson's definition of non-arbitrariness, they are non-arbitrary in both inclusion and exclusion with respect to the other categories, when based on contemporaneous animals.

The import of this criterion of adequacy is that organisms which are grouped together on the basis of their dissimilarities and similarities alone, are indicative of biologically significant continuity between taxa.

(2) Criteria related to the different kinds of degrees of affinity involved in phylogeny. Here, consistency with phylogeny is a primary criterion, while secondary criteria are specified by degrees of similarity and dissimilarity in homologous and, to some extent, parallel characters. These secondary criteria are derived from relative divergence and the sum of dissimilarities in all the characters studied, and apply to the ranking of taxa already recognised rather than to their recognition in the first place. These criteria are an expression of the inclination that taxa of the same rank should be of about the same 'size'.

(3) Criteria related to the relative antiquity of taxa. In evolutionary classification every taxon T_{j+1} is at least as old as any of its included T_{js} (usually species),

and as a rule with possible but certainly rare exceptions, each T_{j+1} is older than the average age of its T_{js}. This is seen as a natural consequence of the nature of phylogeny and the arrangement of evolutionary taxa in a hierarchic system, because with fossil taxa, if taxa of the same rank should be of the same age, then ranking becomes impracticable.

Finally, the recognition of higher taxa, for Simpson, is also based on practical criteria. Higher taxa are defined in terms of phylogenetic patterns:

> A higher category is one such that a member taxon includes either two or more separate (specific) lineages or a segment (gens) of a single lineage long enough to run through two or more successional species.[19]

Their recognition is based on degrees of separation, i.e. gaps, the amount of divergence, and the multiplicity of lower taxa. Fig. 5 exemplifies the relation between higher taxa and their recognition.

Simpson's procedure for ranking is best exemplified in his 1963[21] example of the evolution of the Hominoidea. For nine genera Simpson provided the following classification.

SUPERFAMILY HOMINOIDEA

    FAMILY Pongidae

        SUBFAMILY Hylobatinae

            Pliopithecus

            Hylobates

        SUBFAMILY Dryopithecinae

            Ramapithecus

        SUBFAMILY Ponginae

            Pongo

            Pan

    FAMILY Hominidae

        Australopithecus

        Homo

    FAMILY Oreopithecidae

        Apidium

        Oreopithecus

Fig. 6 shows the accompanying phylogeny. The *Apidium* and *Oreopithecus* lineage is ranked as a family 'because of its ancient separation plus its marked divergence from any other group now usually given family rank.'[22]

Hylobates and *Pliopithecus* are classified with *Pongo* and *Pan* because all of them share a common ancestry and the evolutionary divergence between the *Hylobates* lineage and *Pan* and *Pongo* is much less than for either *Homo* or *Oreopithecus*.

Simpson classifies *Pan* and *Pongo*, and elevates *Homo* to separate familial rank because ranking according to branching points is misleading. 'It might necessitate, for instance, the inclusion of the African apes (*Pan*) in the family Hominidae and their exclusion from the family Pongidae.'[23] Consistency must be upheld at all times.

Having set out the differing attitudes between Mayr and Simpson, I now want to examine in detail their respective claims in the context of theoretical information, for it is in their differing methodologies that this issue is highlighted.

The Mayrian view

Mayr is quite explicit in his designation of classifications as theories,[24] such that classifications have the same properties as theories; that is, they exhibit explanatory power (in asserting that a group of organisms grouped together consists of descendants of a common ancestor),[25] predictive power (with respect to the assignment of newly discovered species and the pattern of variation of previously unused characters), heuristic import[26] (stimulating the generation of hypotheses), and provisionality (classifications can be modified in the light of new discoveries). It is to be expected, then, that classifications must reflect the underlying causes of organic change by functioning as theoretical statements grounded in evolutionary theory. So, all classifications must contain theoretical predicates.

Fig. 5 The relation between grades and taxa. H – Hominoid grade; A – Families, B – Superfamilies, C – Suborders, D – Orders.[20]

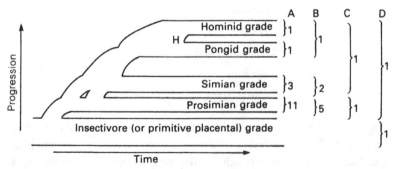

But as we saw in Chapter 1, classifications do not function as theories in this sense. As if in anticipation of this difficulty, Mayr also utilises the second sense of classifications as theories – a shorthand for the conveyancing of information, as described above. Unfortunately, Mayr is never very explicit over which sense he implies at any instance.

In effect, Mayr is attempting to unite the process of classification and the subsequent process of generating hypotheses about evolution. Yet, in *Principles of Systematic Zoology*, it is not made clear how to make such a combination. In what follows, I want to examine the tension that exists in Mayr's thinking between the storage and retrieval of information in classifications, and the explanatory import of classifications. Necessarily, any discussion of the former aspect requires a discussion of the notion of genetic affinity.

Genetic affinity as a measure of evolutionary relationship

For Mayr, the aim of classification is to represent the genetic relationships of organisms. Such relationships are inferred through morphological groupings which should correspond to genetic similarities and differences, so that the constructed groups represent a true reflection of the relationships that exist among genetic programmes of groups, or genotypes of the organisms concerned. At first sight it appears that Mayr is appealing to genetic relationships that the highly trained taxonomist is good at judging, through the weighting of characters. This is plausible enough, but the precise meaning of genetic affinity requires clarification, since there is the possibility of confusion given that the term 'genetic' can have two different meanings, implying either 'gene' or 'origin'. With respect to this double play on 'genetic', I will take genetic similarity to imply similarity of geno-

Fig. 6 Simpson's (1963) view of Hominoid phylogeny.

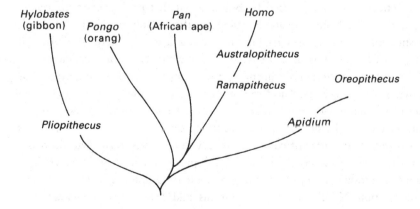

type. The clarification of this problem is largely an empirical issue based on empirical indicators of genetics and types of relationship.

If Mayr is interpreted as implying that 'genetic' means genotype[27] then, in the words of Brundin, 'genetic difference is nothing more than phenetic difference'.[28] The roots of this interpretation are not difficult to find, given what Mayr has said on the subject. Indeed, evolutionary systematists agree with pheneticists in

> grouping by a largely phenetic approach and differ from them in the weighting of characters.[29]

They express similarity relationships rather than kinship relationships:

> It is no longer legitimate to express relationships in terms of genealogy. The amount of genetic similarity now becomes the dominant consideration for the biologist.

> When the evolutionary taxonomist speaks of the relationship of various taxa, he is quite right in thinking in terms of genetic similarity, rather than in terms of genealogy.[30]

But, Mayr never defines his concept of genetic similarity and does not make the genetic systems of taxa the subject of investigation. Genetic similarity is established in other ways. His estimates are phenetic because they are based on 'weighted morphological similarity'. Genetic relationships can only be measured by phenetic similarity.

But, this interpretation of Mayr is wrong for several reasons. Firstly, there are better ways of expressing genotype similarity than Mayr does. Despite the arrival of techniques for measuring genetic similarity in the 1970s, Mayr does not refer to them, preferring instead, his notion of taxonomic weighting. Secondly, and more importantly, it is not clear that similarity between genotypes relates to recency of common ancestry. This is a point that pheneticists have emphasised by pointing out that phylogenetic inferences can be derived from estimates by using specific assumptions. Thirdly, Mayr's notion of weighting is fallacious in this context of genetic similarity. It implies a form of phylogenetic weighting in which taxonomic characters are differentially weighted according to their supposed relative importance as phylogenetic indicators. This form of character weighting presupposes character rarity as a phylogenetic indicator. This is a complete fallacy because the measure of similarity becomes indefinite. In searching for rare characters, as measures of similarity of common characters increases, rarity will decrease. Finally, there is the crucial role of developmental 'currency' in Mayr: how does the product of unpacking the genome in developmental terms reflect its composition? Nothing is known about this, and it can only be concluded that

Mayr fails to accurately represent the totality of the genotype, despite his claims to the contrary. The sense of inferred genetic similarity employed here can only be understood in a restricted sense.

The other interpretation of Mayr, which is most prevalent amongst cladists such as Patterson[31] and Wiley,[32] is that weighted phenetic similarity is a good indicator of phylogenetic relationship. Mayr believes that genealogical relations and genetic relations are not necessarily concordant;

> Let me explain. Since a person receives half of his chromosomes from his father, and his child again receives half of his chromosomes from him, it is correct to say that a person is genetically as closely related to his father as to his child. The percentage of shared genes (genetic relationships), however, becomes quite unpredictable owing to the vagaries of crossing over and of the random distribution of homologous chromosomes during meiosis, when it comes to collateral relatives (siblings, cousins) and to more distant descendants (grandparents and grandchildren etc.). Two first cousins (even brothers for that matter) could have 100 times more genes in common with each other than they share with a third cousin (or brother) (among the loci variable in that population). The more generations are involved, the greater becomes the discrepency between genealogical kinship and similarity of genotype. . .[33]

Wiley criticises this analogy because Mayr 'equates the vagaries of genetic discorrespondence between genealogically related organisms within Mendelian populations with supposed discorrespondence between populations and species.'[34] Within populations genealogical relationships between individuals are subject to the Mendelian process of independent assortment, recombination, and random mating so that genetic similarity may not correlate with genealogical relationship. However, between populations genetic relationships work on a different basis, e.g. the Hardy–Weinberg equilibrium, and the various evolutionary forces which act on that equilibrium, so that Mayr's analogy is misplaced; population genetics is very different from Mendelian genetics.

Practical examples have also added weight to the claim that 'genetic equals genealogical'. Mayr originally argued that in recent Hominoidea,[35] the Pongidae (which includes *Hylobates*, *Pongo*, and *Gorilla*) were genetically more similar to each other than any of them was to *Homo*. However, sampling the genotypes of hominoids through chromosome banding, immunology, electrophoresis, protein sequencing and DNA hybridisation,[36] shows that *Pan* and *Gorilla* are genetically more closely related to *Homo* than to either *Pongo* or *Hylobates*.[37] A further example comes from Mayr's comparison of birds, crocodiles and lizards which shows birds to be more genetically similar to crocodiles in terms of actual loci than to other reptiles; this similarity between crocodiles and birds is also greater than that expressed between crocodiles and other reptiles. As Gorman *et al.* observe

> Every antiserum [of *Alligator*] gave weak reactions in immunodiffusion and microcomplement fixation with bird albumin. . . Lizard and turtle albumin failed to react with anti-*Alligator* albumin.[38]

Another example of genetic similarity as a good indicator of genealogical relationship is given by Patterson.[39] Using a shark, a teleost and bird or mammal in place of the lizard, crocodile and bird system (Fig. 7), Patterson shows that the bird/mammal and teleost proportion of genotype outweighs that shared by the teleost and shark.

Under this interpretation of inferred genetic similarity, Mayr's notion of weighting becomes vacuous, since any indicator of genealogical relationship necessarily precludes any talk of weighting in a phyletic context. There is also a confusion over estimates of weighting[41] based on primitiveness and conservatism. The two are not necessarily the same, contrary to what Mayr may have implied.[42] A character is said to be primitive in a group of organisms if it was present in the common ancestor of the group. The degree of primitiveness will depend upon the time since divergence (evolutionary age) of the group of organisms. A character is conservative if it occurs in all descendants of the population in which it arose. A conservative character may be of recent or ancient origin.

Fig. 7 Genetic and genealogical relationships.[40] Percentage shared genotype estimated from the amino acid sequence of myoglobin, and alpha and beta haemoglobin: (1) for the shark (*Heterodontus*) and teleost (*Cyprinus* – carp) the percentage of codons shared with respect to *Homo sapiens* is 9.9%, and with respect to *Gallus* (chicken) is 8.8%; (2) for the teleost (*Cyprinus*) and bird/mammal the percentage of shared codons is 19% (carp/man) or 19.5% (carp/chicken).

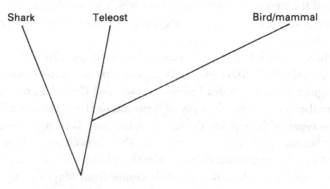

Shark   Teleost   Bird/mammal

However, I do not think that either interpretation of inferred genetic similarity does justice to what Mayr intends. As I mentioned earlier, Mayr believes that genealogical and genetic relations are not necessarily concordant. Such an attitude implies that Mayr is equivocating, and that he takes a combination of both recency of common ancestry and similarity of genotype. It is then possible to argue that for some combination of these two indicators the true evolutionary relationship is elucidated. Unfortunately, there is a fatal flaw in this: there are so many variables possible in combining these two factors, that any view of inferred genetic similarity as an indicator of evolutionary relationship is incoherent. In addition, the possible alternatives in interpretation that I have presented above contain obvious falsehoods. Bock argues that the degree of genetical similarity is ascertained by the degree of similarity of phenotypical features shared in common. Thus greater phenotypical similarity implies greater genetical similarity. But not only are attempts to infer the genotype from the phenotype difficult, but also cases have been found in which there is morphological convergence with extreme genetic difference.[43]

In conclusion, Mayr's inferred genetic similarity is a bad indicator of evolution,[44] irrespective of the interpretation given to it. With this conclusion in mind, I now want to turn to the problem of representing evolutionary assumptions in a classification.

The representation of evolutionary assumptions in a classification

Of importance here are the rules for the representation of evolutionary assumptions in a classification. What rules are required in constructing a classification indirectly from a phylogenetic tree? I will concentrate on two aspects – firstly, the importance of ranking with respect to the storage and retrieval of information in a classification; and secondly, the question of stability in classifications with respect to the mode and tempo of the underlying evolutionary process. In both cases, it will be clear that the methods are not concordant with the purposes stipulated by Mayr.

For Mayr, a classification must have two objectives: grouping close relatives (determined on the basis of inferred genetic similarity and various ranking criteria) and facilitating the storage and retrieval of information.

> The complex relationship of species with each other and the varying rates of branching and divergence must be translated into a system of taxa, conveniently ranked in the appropriate categories in such a way that it forms as efficient as possible a system of information storage.[45]

Mayr also claims that his classifications contain greater information content of a specific evolutionary nature:

> . . . cladists seem to think that they have to make a choice, in the delimitation of taxa, between basing them *either* on branching points or on the degrees of evolutionary divergence. They fail to appreciate the added amount of information obtained by utilising *both* sources of evidence.[46]

Now, in arguing that classifications are theories (in the sense of maximising information content), Mayr is asserting that they can convey simultaneously, knowledge of genealogy, genetic similarity, adaptive divergence etc., through the judicious assignment of rank. Such information is stored in the classification. However, Mayr fails to provide any rules or criteria for retrieving that information from his classification. For instance, the question of whether to place man in the same family as *Pan* (the African apes) or not, is resolved by assigning man to a separate family because through the familial categorical rank the degree of difference between *Homo* and its relatives is indicated. But the difficulty with attempting to use categorical rank to convey information about degree of distinction between taxa is that in order for it to work, the categorical level of separation between the two taxa must generally be interpreted in terms of degree of difference. In the above example, *Homo* is separated from *Hylobates* at the same categorical level as that which *Homo* is separated from *Pan*, conveying the 'information' that *Homo* is equally distinct from *Pan* and *Hylobates*. Yet, *Homo* is admitted to be much more similar to *Pan* than it is to *Hylobates*. Thus, basing categorical rank on degree of difference distorts the information that it is intended to express. Classifications cannot store and retrieve the information that Mayr requires.

A possible rejoinder to this is expressed by Bock:

> Because the approach of evolutionary classification does not insist on a one-to-one correspondence of the classification and the phylogeny, separate sets of hypotheses about groups are needed to cover both aspects of relationships. And it is necessary at the completion of an evolutionary classificatory analysis to represent these conclusions in a formal classification and in a phylogeny. Many workers omit the phylogeny; this omission causes problems for others who may need the exact phylogeny for their studies and yet cannot obtain this information from the classification.[47]

Classifications such as Mayr's are supposed to contain a measure of morphological and adaptive divergence, but that such information can only be retrieved with the aid of a phylogenetic diagram. It is therefore misleading for Mayr to say that classifications can store knowledge of genealogy, adaptive divergence etc., when such information cannot be retrieved without a phylogenetic diagram.

The reason for this failure stems from the definition of relationship utilised – inferred genetic similarity as determined both by distance from branching points and subsequent rate of divergence. In subsequently deciding the rank of taxa, Mayr, like Simpson, advocates a combination of gradistic and cladistic factors. But for Mayr, this is a compromise

> . . . in the derogatory sense of the term. Similarities and dissimilarities calculated with phylogenetic weighting will be a function both of the relative times of branching in evolution and of the subsequent amounts of phenetic divergence in populations. They will be unknown and variable functions of phenetic similarity and time since branching in evolution.[48]

In other words, a classification based upon recency of common ancestry and degree of genetic divergence will, in attempting to convey both kinds of information, convey no information. Mayr fails to live up to the standards that he sets himself because his aims are not concordant with his methods.

I now want to examine the question of stability in classification with respect to the mode and tempo of the underlying evolutionary process. More specifically, this concerns the status of 'gaps' in determining a classificatory scheme. Such gaps are usually expressed morphologically.[49] Morphological gaps play a central role in the grouping of close relatives and in the assignment of rank. In the former case, Ashlock[50] has documented Mayr's recommendation that the use of the gap be an inverse ratio to the size of the taxon.[51]

> In other words, the more species in a species group, the smaller the gap needed to recognise it as a separate taxon, and the smaller the species group the larger the gap needed to recognise it.[52]

In the latter case, if a sufficient morphological gap exists between two sister groups, then they may be separated and placed in rank-coordinate, higher taxa. For example, the large morphological gap between the Aves and Crocodilia is expressed through the elevation of Aves to the class category, while the Crocodilia are relegated to the Reptilia as an order.

The most important point about such rules for expressing morphological gaps is that they involve practical judgements. With Mayr's emphasis on classifications as theories, there is no place for pragmatic considerations because the method does not fulfil the aim. Indeed, the smuggling in of pragmatic criteria by Mayr is totally at odds with his other methodological considerations such as ranking on the basis of phylogenetic affinity.

Furthermore, there are very real problems over the evolutionary status of gaps, and the particular assumptions that are contained in the concept of a gap. Two perspectives are possible.

(A)   Gradualism

If it is assumed that evolution is gradual,[53] then the morphological gaps must be deemed artificial; that is, they are the result of sampling errors. Within this framework, it should be possible to eliminate gaps between closely related (genealogically, that is) groups on the basis of further sampling, although in practice this is not possible due to the extinction of intermediate species,[54] or the nature of the material at hand (e.g., imperfect preservation in the fossil record). However, as the gaps disappear, then the basic ranking rationale based on gaps becomes meaningless.[55] Under the assumption of gradualism, the problem with gaps is that the theory is not amenable to the taxonomic application.

(B)   Saltation

If it is assumed that rapid morphological divergence of species from their ancestral morphology has taken place, then any morphological gap between two sister species is a potential absolute gap. Some morphological gaps are real and not the byproduct of sampling error. However, the significance of the gap depends on the type of saltation assumed. Two types are possible. In microsaltation[56] only small gaps are produced (Fig. 8), and it is new species that are produced, not new classes or orders. Potential error arises over missing species, and thus missing speciation events which couple the microsaltation with lineage divergence. No form of microsaltation, when coupled with speciation, allows for sequential anagenetic saltation. Indeed, Eldredge & Gould reject anagenesis as a major evolutionary force. But biases may arise in cases where large gaps are observed. Such gaps may be due to sampling error since the expected gap size is no larger than those observed between relatively close branching points. The result for evolutionary classification is the same – the theory does not complement the classification.

In macrosaltation,[58] if it is believed that large-scale saltations do occur in nature, then some morphological gaps are absolute gaps. There are major morphological changes of a form as observed between major clades. In cases where large morphological gaps may reasonably be ascribed to macrosaltations, the resulting groups are usually accorded equal rank, e.g. class. With respect to evolutionary systematics, macrosaltation has never been invoked as a causal factor in the formation of large gaps. It follows, for

Mayr, that almost all gaps are artificial (even if a liberal reading is taken, such that the synthetic theory does not entail that evolution *must* be gradual or all pervasively adaptational[59]). This means that the stability of the resultant classification is impaired,[60] because as the gaps are filled in with either newly discovered fossils or Recent taxa, the justification for the ranking criterion diminishes.

Because gap recognition constitutes evidence either for gradualism or saltation, Mayr is misleading in attempting to represent a particular view of evolution, i.e. gradualism, in a classification. Classifications cannot accurately represent evolutionary processes. More importantly, Mayr is empirically wrong: classifications obtained using his methods will not mirror recency of common ancestry and anagenetic change. It follows that the information stored in a classification does not give adequate retrieval in the manner that Mayr intends. The aim and methods are discordant.

Given that there are limits to the storage and retrieval of information in Mayr's system, we are now in a position to see how this affects the explana-

Fig. 8 Punctuated equilibria and phyletic gradualism.[57] (a) Punctuated equilibrium – all change is concentrated into the speciation event. (b) Gradualism – change is more or less continuous. (c) Variation in punctuated patterns in which (i) the parent species is extant, and (ii) the parent is extinct.

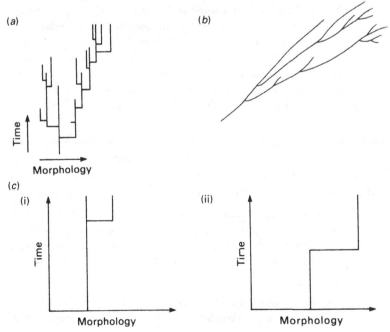

tory aspect of classifications, and whether increased information content is more likely to lead to the uncovering of an explanation.

Species, higher taxa and explanation

For Mayr, because classifications are explanatory, they are *ipso facto* theories:

> Taxa delimited in such a way as to coincide with phylogenetic groups (lineages) are apt not only to have the greatest number of joint attributes, but at the same time to have an explanatory basis for their existence.[61]

But, in Chapter 1, we saw that classifications as theories failed on the grounds that information is classified monothetically, and the more information stored does not overrule the criterion of classification. Now, Mayr's views on explanation and theory can be saved from the problems of monothetic criteria by the use of explanatory essences, since explanatory essences are compatible with a 'no fixed criterion' view.

The notion of explanatory essence has been discussed in recent times by Putnam,[62] Kripke[63] and Wiggins.[64] Platts[65] also gives a clear exposition of the ideas involved and I will base my account on him. Crucial to the discussion of explanatory essences is the definition of terms. An expression is said to be a natural kind word if the extension of the expression constitutes the membership of the natural kind.

> A class of objects in this world is the actual membership of the natural kind *o* only if a correct answer to the question 'what are these?' asked of the members of that class is '*os*'. Truth does not suffice for correctness. Rather, the answer '*os*' is correct if, and only if, it is potentially a sufficient explanatory characterization of the members of that class, where this explanatory potential stems from law-based explanations of the properties of the members of that class which could truly be given.[66]

What is of importance in a class being a natural kind class is that there are explanations (of law-like form) which can account for the nature of each member of that class. Thus, the important physical properties of a natural kind class are those important for explanatory purposes. As Wiggins explains

> x is an f (horse, cypress-tree, orange, caddis-fly) . . . if and only if, given good exemplars of the kind. . . , the most explanatory and comprehensive true theoretical description of the kind that the exemplars exemplify would group x along-side these examplars.[67]

In the case of tigers,

> The class of actual tigers is the actual membership of the natural kind *tiger* because, when asked 'what are these?' of the members of that

class, the answer 'tigers' serves potentially to explain sufficiently what these objects are, their properties and their doings, by drawing upon the law-based explanatory credit to be met by the discovery of the true tiger-theory. That theory, when discovered, will serve through its constituent laws to account in the appropriate way for sufficient of those properties and doings.[68]

Explanatory essentialism is also an attractive view for species, for in order to belong to a species, there must be a hidden property which explains various characteristics of species. For Mayr, this property is the unity of the genotype. Unity of the genotype implies that

> . . . the genotype consists of a number of co-adapted gene complexes and that even the whole gene pool of a population as a whole is well integrated and co-adapted.[69]

Ignoring both stochastic processes and ecotypic selection in the determination of allele frequencies, only the unity of the genotype can be invoked to explain why different sets of alleles will be favoured in every species. The speciation process necessarily involves a thorough reorganisation of the genotype.

Mayr himself does not discuss explanatory essentialism, but it represents a possible way in which his claim of classifications being theories can be saved. However, for it to work, explanatory essentialism must apply to the units of taxonomy – species and higher taxa. In the case of species, explanatory essentialism is plausible since, instead of species being defined by genotypes, species can be defined in terms of genotype unity that leads to an explanation of other characters to be had. Some structural factor explains various properties of species, in that characteristics of these relatively stable gene complexes also explain their behaviour.

In the case of higher taxa, Mayr invokes genotype unity to explain various properties, such as the conservative nature of the *Bauplan* of the major animal types, e.g. insects are hexapods, land vertebrates are tetrapods.[70] Within the 24 phyla of Recent animals, the basic features of the structural plan of these phyla is derived from that of the founders of these types. Unity of the genotype helps to explain why a Bauplan (e.g. metamerism), once established and properly consolidated in the genotype, remains remarkably stable regardless of subsequent developments. Unity of the genotype is also invoked to explain ontogenetic trends.

But, whatever the merits of explanatory essentialism for species, higher taxa cannot be given stable gene complexes in the same sense. For Mayr, species are related to ecological niches, and while unity of the genotype may be invoked to explain the more conservative aspects of speciation, such as the seeming uniformity of a species over wide areas in spite of drastic environmental differences, this type of explanation cannot be used in explaining properties of higher taxa. In higher taxa the ecological input is lost, so while species of the world constitute the total actual membership of the natural kind species, they are not merely part of the actual membership of the natural kind 'all taxa' (i.e. species and higher taxa). Although various properties of species can be used to explain various properties of higher taxa, this is not the same as saying that higher taxa manifest explanatory essences. Quite the contrary, since the appeal to explanatory essences in higher taxa leads to disjunctive explanations for each species because genotype unity has to break up at, say, the ordinal level. For example, in the order Carnivora, dogs are members of the infraorder Arctoidea, while cats belong to the Aeluroidea.[71] In giving an explanation of this at the ordinal level, there is a disjunctive explanation at the infraorder level, requiring a breakup of genotype unity at the ordinal level.

While explanatory essentialism is an attractive view at the species level, it is not possible for higher taxa, implying that a much more pragmatic view has to be taken; classifications cannot function as theories even in the light of explanatory essentialism.

In conclusion, it is clear that there is a lack of conguence between Mayr's methods and aims. Mayr not only presents a view of evolution that is empirically incorrect, but also fails to take account of factors such as the reversibility of evolution and gap formation. Furthermore, his rules for the storage and retrieval of information in classification have a practical emphasis which is not consistent with his claim that classifications are theories. A lack of clarity over the possible meaning of classifications as theories leads Mayr into difficulties. In using classifications as theories for a shorthand of 'maximizing information content', Mayr confuses methodological rules with theoretical assumptions. Rules used for the storage of information cannot be converted into explanations for the justification of his system. Nor are such explanations carried within the classification, highlighting the vacuousness of the view that the more information that is stored in a classification the more likely a correct explanation will be forthcoming. Conclusively, Mayr's analogy of a classification with a theory is carried much too far. Ehrlich & Ehrlich[72] say as much in commenting that there is no way of checking such a theory experimentally in a manner comparable with the verification/rejection of other scientific theories.

The Simpsonian view

I do not intend to say very much concerning Simpson's views on classification, because he does not push up against the limits of information storage and retrieval in classifications or the problem of explanation. Simpson does not talk of classifications as theories and is therefore not open to the charge of inconsistency between methods and aims. His concern is over much more practical issues with an emphasis on the commonsense suspension of theoretical issues that are regarded as too rigid for practical use.

However, Simpson's discussion of consistency between classification and phylogeny does imply some kind of theoretical status for classifications. This view is further substantiated by his emphasis on monophyly as the main criterion of classification. I do not think, though, that such claims are fatal for two reasons; first, because Simpson does not advocate a strict correspondence between classification and phylogeny, there is room for other, more practical criteria, such as the size of taxa in rank determination, to be utilised. Simpson's classifications are not based on one criterion alone. Second, and more importantly, Simpson is concerned with practical problems, such as the inclusion of fossils. In this respect his aims and methods are sensible and realistic (the full force of this point will become much clearer after my discussion of Hennig's system). Thus, while there may be a lack of specificity in what can and cannot be conveniently represented in classifications, this is to a large extent an inherent aspect of the classificatory activity (i.e. combining complexity of information with accuracy).

Such defences may appear inadequate, but they can only be fully appreciated in the light of problems that other taxonomic schools come up against. For instance, in the case of cladistics, it will be argued that it is difficult to reject outright the use of fossils.

Finally, I want to say a little about the differences between Mayr and Simpson. I have argued, in the case of Mayr, that classifications must be based on cladistic and gradistic elements, and that these elements are mirrored in his use of inferred genetic similarity.[73] Such a method is incapable of detecting parallelisms, and more conclusively, does not give an accurate picture of evolution. Simpson, on the other hand, combines cladistic and gradistic elements through the use of fossils. One advantage here is that parallelism can be more easily detected, giving a more realistic picture of evolution. Of course, both Mayr and Simpson confront similar problems in the storage and retrieval of information, but for the latter this is not such a fatal flaw because he is not making unrealistic claims over the status of classifications.

Neontological versus paleontological approaches

In conclusion to this discussion of evolutionary systematics, the differing attitudes present in Mayr and Simpson can be explained in terms of the neontological and paleontological approaches respectively. This distinction is important because it is an issue that is taken up in earnest by the cladists. These two approaches become evident in the following question: should evolutionary phylogeny attempt to describe the inter-relationships of living taxa alone, with fossils acting as possible, though not necessary, means of illuminating these relationships (although fossils are irrelevant in themselves)? Or, should evolutionary phylogeny study the interrelationships of the diversity of life throughout the Phanerozoic, with living taxa being the surviving end points that can throw additional information (through soft anatomy, biochemistry, cytology etc.) on the pattern of the branching tree? Simpson is quite explicit in his choice:

> Neontologists may reveal what happens to a hundred rats in the course of ten years under fixed and simple conditions, but not what happened to a billion rats in the course of ten million years under the fluctuating conditions of earth history. Obviously the latter problem is more important.[74]

Newell is equally insistent in his conclusion that

> hypothetical phylogenies based soley on living genera and species cannot express the true relationships. In order to understand the ancestry of, and connections between living genera and families it is necessary to know the fossil record.[75]

In contrast, neontologists such as Mayr do not see any need to know the fossil record.

In applying the neontological and paleontological approaches to classification construction, there are important ramifications for the storage and retrieval of such information. Of course, all evolutionary systematists acknowledge that classifications are compromises between the fundamentals of evolutionary theory and their representation in a classification, and that classifications cannot be entirely cladistic (based on clades) or gradistic (based on grades). But here the similarities between Mayr and Simpson end.

Because Simpson believes that a study of fossils is integral to establishing evolutionary branching sequences, then the problem of expressing relationships between fossil and Recent organisms in a classification can be accomplished only by relaxing the criterion of consistency to minimal monophyly. With strict monophyly, the incorporation of fossils would lead to the piling up of categories, and cumbersome classifications. For Mayr, the definition of evolutionary relationship in terms of inferred genetic similarity means that fossils are ruled out on principle, though not necessarily in practice (since Mayr does not use information from genetic sequencing, implying that inferred genetic similarity can be estimated from fossils). In the absence of fossils, minimal monophyly is not necessary, but there remains instead the difficulty of combining estimates of parallelism and convergence with some degree of monophyly. Since Mayr incorporates grade taxa, a balance between detecting parallelisms and accuracy of branching sequences must be maintained, and this can only be done through careful specification of monophyly in the light of undetected convergence. But, because grades are composed of independent lineages progressing by parallel development through sequences of adaptive zones, the use of inferred genetic similarity will not necessary distinguish between such parallelisms and convergences. The detection of parallelism and convergence can be accomplished more satisfactorily through the fossil record.

In conclusion, the methods of Mayr and Simpson exemplify a tension that exists between either the weakening of monophyly and the inclusion of fossils in a classification, or the possible loss of information through the rejection of fossils (e.g. undetected convergence) combined with increased ease of storing phylogenetic information in a classification without the piling up of categories. These are problems that will also be confronted in Hennig's work.

Within the writings of both Mayr and Simpson, a general methodology for inferring phylogenies is distinctly lacking and any discussion of this topic is kept to a minimum. It is stressed, instead, that

> because of the subjective nature of the problem it is difficult to lay
> down any hard and fast procedures for attaining satisfactory results.[76]

This attitude highlights the status of evolutionary systematics as a largely intuitive approach in which

> there is an element of circularity, the inferences which are to be based
> upon the result of the classificatory method being tangled up with the
> methods of classification themselves.[77]

Given this view of classification and its relation to phylogeny, it will come as no surprise that the more numerical approaches had little option but to reject the use of phylogeny altogether in the construction of classifications

3

Phylogenetic cladistics and theoretical information

Since the publication of Hennig's *Phylogenetic Cladistics* in 1966, cladistics has undergone many changes such that a diversity of approaches can be made out in the literature. In this section I want to concentrate on the present-day views of phylogenetic cladists such as Cracraft, Eldredge[1] and Wiley.[2] At the same time I will outline the basic views of Hennig, so that the emphasis is on what the latter-day phylogenetic cladists draw from Hennig, and on how they modify it. In this way we will be better prepared to examine the move to transformed cladistics.

For the most part it is assumed that cladistics arose with Hennig – the two are seen to be synonymous. However, this is not strictly true because many evolutionary taxonomists before the arrival of Hennigian cladistics, had decidedly cladistic attitudes. Indeed the discussion of 'horizontal' and 'vertical' classifications[3] anticipated some of the issues raised by Hennig's cladism, and such discussions had their roots in the issues over cladism or gradism. For example, the Swedish school of paleontologists, headed by Stensiö,[4] Jarvik[5] and Säve-Söderbergh[6] had strong cladistic leanings. Furthermore, there is some evidence that the Italian zoologist Daniele Rosa (1857–1944) invented, or at least anticipated, cladistics.[7] Historically speaking, then, Hennig did not invent cladistics.[8] It had been independently anticipated by at least two sources. However, Hennig's contribution is still substantial and his originality stems from an attempt to detect patterns in present-day organisms.

Hennig's strict belief in evolution is clearly reflected in his aims. Given that evolution has produced a natural system of relationships among organisms through genealogical descent and modification, then it is the job of systematists[9] to discover these relationships. This system of relationships can be communicated to other biologists by using a language, a classification, that is as natural and as informative as the relationships that are discovered.

Like evolutionary systematists, Hennig believed that evolutionary theory should play a central role in taxonomy, and that biological classification should have a systematic relation to phylogeny. But here the similarities end. Hennig stipulated that this relation between phylogeny and classification should be a direct one-to-one correspondence, and that the aspect of phylogeny which could be represented explicitly was the order of branching. This latter point is reflected in the derivation of the term 'cladistics', which is taken from the greek 'KLADOS'[10] meaning a young shoot or branch.[11] Hennig gave two reasons for representing the order of branching in classifications; firstly, the Linnaean hierarchy, as a list of indented names, lends itself more naturally to expressing discontinuous variation rather than continuous variation. Secondly, the order of branching can be ascertained with sufficient certainty to warrant the inclusion of such information in a classification, while other aspects of phylogeny, such as anagenetic change cannot be. Thus Hennig advocated the phylogenetic system because it had certain logical attributes and priorities that made it the best choice for a general reference system. A classification was ideally suited to carrying explicitly theoretical information about evolution. In this respect, Hennig attempted to strike a balance between how much information could be stored in a classification and the subsequent complexity of the rules for representing such information. He utilised strict methodological rules in the construction of cladistic classifications, thereby purging the phylogenetic system of any 'uncertain' elements. Hennig, therefore, stands midway between the intuitive and numerical approaches (see Chapter 4).

It is important to note that all phylogenetic cladists adopt these aims. Thus Wiley[12] re-emphasises the point that the relationships leading to the cohesion of living and extinct organisms are genealogical ('blood') relationships, and that such genealogical relationships among populations and species may be recovered by searching for particular characters which document these relationships. Consequently, the best general classification of organisms is the one that exactly reflects the genealogical relationships amongst those organisms. Differences between Hennig and later phylogenetic cladists arise over what information is to be stored in classifications, and the rules used in achieving these ends. There is a concordance over aims, but the methods differ.

Hennig's phylogenetic method has two components.

(1) Cladistic analysis, which incorporates rules for discerning cladistic relations and the rules for representing them in cladograms.[13]

(2) The construction of a classification from the cladogram.

Most of the developments since Hennig have concentrated on cladistic analysis, with a greater stress on methodology, empirical grounding and testability.

Cladistic analysis

Cladists argue that by and large all a taxonomist has to go on are the traits of the specimens before him. On the basis of a comparison of organisms, perceived similarities are utilised to choose those characters that will be used for the more detailed comparisons leading to the construction of a cladogram. Central to this endeavour is the recognition of shared derived characters.

Derived characters are evolutionary novelties, representing the inherited changes of previously existing characters to new characters. Phylogenetic cladists, like Hennig, express the recognition of shared derived characters in terms of the concept of synapomorphy:[14]

> The condition of sharing a derived character or a later stage in the transformation sequence of a derived character-state is termed a synapomorphy.[15]

In Fig. 9, *b'* is a diagnostic character unique to taxa X and Y, and is termed a synapomorphy. In contrast, *a* is a primitive character state[16] common to taxa X and Y, and is called a symplesiomorphy. For example, feathers and a furcula are synapomorphies unique to *Archaeopteryx* and birds, while teeth and a long tail in *Archaeopteryx* are symplesiomorphies – they are also common to reptiles and fish.

It is worth noting at this point that the recognition of shared derived characters involves an extreme form of character weighting with respect to

Fig. 9 Three taxa X, Y and Z possess two characters, in either a primitive state (*a* and *b*) or in a derived state (*a'* and *b'*).

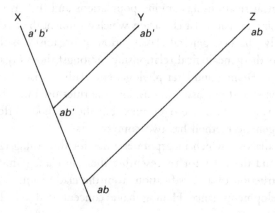

underlying assumptions about evolution, specifically the branching pattern. Only those factors which are evolutionary novelties express a relationship in terms of recency of common ancestry. Any information concerning the evolutionary origins of taxa is rejected. This means that ancestor–descendant relationships cannot be expressed by phylogenetic cladists since synapomorphies are shared similarities inherited from ancestors more remote than the immediate common ancestor (Fig. 10).

All phylogenetic cladists argue that synapomorphies define nested sets within a cladogram, and that the recognition of such sets is specified by monophyly. Hennig originally argued that every taxon should be monophyletic in the sense that

> it is a group of species descended from a single ('stem') species and which includes all species descended from this stem species.[17]

So all, and only all those species (both extinct and extant) which are descended from a single species must be included in a single higher taxon.[18] Present-day phylogenetic cladists, such as Eldredge & Cracraft, equate monophyletic groups with natural groups,[19] and argue that the search for natural groups is the search for patterns of synapomorphies. Given that cladistic analysis involves the search for monophyletic groups as evidenced by synapomorphies, all phylogenetic cladists argue that the essential procedure is

> the formulation of hypotheses of synapomorphy and their nested pattern. Subsequent evaluation of hypotheses of synapomorphy and of the cladogram itself rests with the prediction that future synapomorphy statements will continue to be congruent with those of the cladogram.[20]

While the importance of concepts such as monophyly and synapomorphy are accepted, there is variation in methods of character analysis, and in the subsequent procedures for formulating and testing hypotheses of synapomorphy (cladograms). I will document three methods which give a cross-section of those available in phylogenetic reconstruction: Hennig's method, the Wagner method, and compatability analysis.

Fig. 10 The two forms of evolutionary relationship. (*a*) By common ancestry, (*b*) by descent.

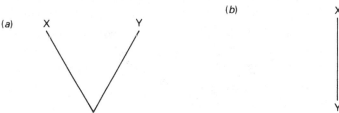

(a)   *Hennig's method*

The distribution of character states is noted (in the example below six characters are scored for five taxa, A–E) and one of two states is assigned to each character – 0 if it is ancestral, I if it is derived. Each character is taken as providing evidence for the existence of one monophyletic group in the phylogeny. Furthermore, it is assumed that each derived state has arisen once only, and that evolution is irreversible – no character in the derived state ever reverts back to the ancestral state. As we shall see, both assumptions are extremely dubious. From the table of characters in Fig. 11, character 1 defines the monophyletic group in which state I arose from state 0. Any other arrangement would require two origins of state I or a reversion of state I to state 0. Similarly for the other characters,

character 2 – group A, C, E

       3   B, D

       4   D

Crucially, character 4 makes no real contribution to the phylogeny, since it is originally assumed that each of the taxa A through E is itself monophyletic.[22] However, if we go on to define the groups specified by characters 5 and 6 (i.e. B, C, E), then there arises an incompatibility.

Fig. 11 Hennig's method for inferring phylogeny.[21] Bars mark the point at which character state changes would occur if the number of changes were minimised. The number of the character that is changing is shown beside each bar.

Taxon	Character					
	1	2	3	4	5	6
A	I	I	0	0	0	0
B	0	0	I	0	I	I
C	I	I	0	0	I	I
D	0	0	I	I	0	0
E	0	I	0	0	I	I

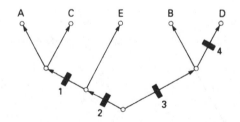

Which combination, i.e. characters 1-4 or characters 5 and 6, gives the true phylogeny? In the face of such contradiction, all Hennig could recommend was to go back and re-examine the characters, in the hope that such incompatibilities could be resolved. Furthermore, the status of derived states in specifying recency of common ancestry is questionable due to the problem of reversals. In the Hawaiian *Drosophila* complex,[23] phylogenetic relationships are inferred from the sequence of bands in polytene chromosomes, so that chromosome inversions give a branching sequence. Yet, in the genitalia, the use of derived states gives a different branching sequence, possibly implying that the derived states in genitalia have arisen many times. Evolutionary reversals are also well documented in teeth, e.g. dedifferentiation of the carnassial teeth has occurred in several families of Carnivora[24] (bears, procyonids, mustelids and viverids) and in the Agnatha.[25] Clearly, synapomorphies do not necessarily give the correct branching sequence[26] in the light of evolutionary reversals. While there may be a case for arguing that the validity of derived states is dependent upon the level of analysis, this is undercut by difficulties peculiar to each level, and in the relation between each level. Indeed, Hennig appears to include genetic units as morphological units, thereby confusing two separate levels of analysis.

(b)   The Wagner method

This method was designed to give an adequate topology of the phylogeny without two of the restrictive assumptions of Hennig's method, namely the specification of each ancestral character state, and the irreversibility of derived states. To overcome these problems, the Wagner Parsimony method was based on allowing characters to have reversions from state I to state 0, or in allowing state I to arise more than once in each character, such that the best phylogeny is the one which minimises the number of these extra events. However, the Wagner Parsimony method gives an unrooted tree (or 'Network'),[27] so that for each possible placement of the root there corresponds an assignment of ancestral states to the characters. In Fig. 12, if the root was placed at the fork which connects B to the tree, this would imply that the ancestral states in characters *1, 2 & 4* were 0, and the ancestral states in characters *3, 5 & 6* were I. For this reason the position of the root is not indicated in the tree.

In cases where the character states are known at the fork, it is possible to specify ancestral characters as hypothetical ancestors in a way that requires as few state changes as possible.

(c)   Compatibility methods

These methods attempt to make a compromise among the characters, by finding the largest set of characters that are mutually compatible, and using only those characters to build the phylogeny. Compatibility methods were first suggested by Wilson,[28] although LeQuesne[29] provided the first critical step by using the largest set of mutually compatible characters to infer the phylogeny. Estabrook[30] and McMorris[31] built on this by specifying a formal definition, the Pairwise Compatibility Theorem:

> a collection of characters are all pairwise compatible if and only if they are mutually compatible, which means that there is a phylogeny on which all of them could evolve with each state arising no more than once.[32]

This theorem is applicable to characters with two states (with or without a specified ancestral state) and to multistate characters (with ancestors specified). Referring back to the data in Fig. 11, the results of LeQuesne's compatibility method are given in Fig. 13. The first four characters give the largest 'clique' of mutually compatible characters. The next largest clique is given by characters 4, 5 and 6. Thus tree topology is based on the largest clique.

In the methods discussed above, all the phylogenies must be rooted in the construction of a cladogram. Furthermore, each of these methods presupposes an *Additive Model* of evolution,[33] so that the resultant cladogram has an Additive structure. The Additive Model is the basis for any method of phylogenetic analysis which attempts to impose Additive structure on a distance matrix, or which searches for patterns of nested sets of taxa revealed by the possession of character states in common. The Additive Model can be stated thus:

Fig. 12 Phylogeny obtained by Wagner Parsimony Method. Same data as used in Fig. 11. The location of postulated changes are shown by bars with the number of characters marked beside each. The phylogeny is intrinsically unrooted, since the root could be placed anywhere on this tree without affecting the number of changes of state reconstructed.

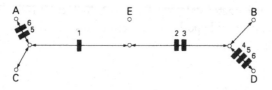

$$d(w,x) + d(y,z) < \max \; [d(w,y) + d(x,z)],$$
$$[d(x,y) + d(w,z)]$$

for all x,y,z.

In Fig. 14, A is always less than B *min*, so that the model assumes evolution to be divergent. Parallelism and convergence are rejected outright. Obviously, this is a view of evolution that is not biologically sound, since both convergence and parallelism are well known evolutionary phenomena. Thus, phylogenetic cladists assume a particular view of evolution, even before they begin to analyse the pattern of nested sets, and its phylogenetic significance. For the phylogenetic cladist it does not matter whether speciation is sympatric or allopatric, saltative or gradual, Darwinian or Lamarckian, just so long as it occurs and is predominantly divergent.

All phylogenetic cladists incorporate a methodology that is more complex than Hennig's. Not only do the results of character analysis lead to cladogram construction, but also these results are the major basis for accepting one cladogram over another (i.e. testing competing cladograms). Character congruency is here regarded as central to all aspects of cladistic analysis because it is based on the fundamental assumption (in phylogenetic cladistics) that the evolutionary process has led to a unique history of life which, in history, can be represented by the one true cladogram.

Eldredge & Cracraft[34] specify two major components in cladistic analysis. Firstly, the development of cladistic hypotheses (cladograms) and secondly, the evaluation of one hypothesis relative to an alternative hypothesis. In the first component, it is the cladogram specifying a pattern of relationships (nested sets) among taxa that represents the hypothesis, such that relation-

Fig. 13 LeQuesne's compatibility method. State changes are marked by bars, the number of the character changing being shown next to each change of state.

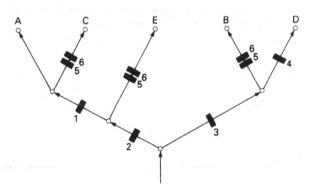

ships are specified in terms of nested sets of synapomorphies. This is some-what different from Hennig's original characterisation of relationship as recency of common ancestry, since the emphasis is now on synapomorphies, and patterns of synapomorphies that are internested. Eldredge & Cracraft also suggest the use of outgroup comparison to overcome the problem of incompatible groups specified by competing derived characters. In attempt-ing to resolve a scheme of synapomorphy within, say, competing three-taxon cladograms (Fig. 15), those character states occuring in other taxa within a larger hypothetical monophyletic group that includes the three-taxon statement as a subset, can be hypothesised to be derived. The method, therefore, maps character state distributions within and without the study group. If taxa A, B and C are members of the larger monophyletic group (A–E), then we can compare A, B and C to outgroups. Those characters found in the outgroups are primitive within the three taxa (A, B, C). In effect, outgroup comparison involves the search for the hierarchical level of character states, i.e. synapomorphies.

In the second component, competing cladistic hypotheses are evaluated using synapomorphies (Fig. 16). The chosen hypothesis is the one that exhibits the greatest congruency of characters, because it is assumed that the unique historical pattern should exhibit maximal congruence of the nested symapomorphies. Phylogenetic cladists justify this view on the basis of parsi-mony, in that the criterion of parsimony specifies our acceptance of the least rejected hypothesis.

Further elaboration of the phylogenetic method also occurs in the con-struction and testing of trees as being a distinct procedure from cladogram construction. Hennig never recognised this distinction. Phylogenetic cladists regard a cladogram as a diagram of entities specified by the inferred historical connection between the entities on the basis of synapomorphies. In contrast,

Fig. 14 The Additive Model. Based on a labelled, unrooted tree shape derived from the differences in the states at the tips of the trees.

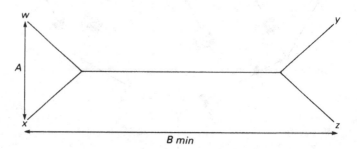

a phylogenetic tree is 'a branching diagram portraying the hypothetical genealogical ties and sequence of historical events linking individual organisms,
populations or taxa'.[37] For any given cladogram, a number of phylogenetic
trees can be derived, the number dependent upon the number of taxa contained within the cladogram (see Fig. 17).[39] But, because trees also include
ancestor–descendant relations, and episodes of multiple speciation and reticulate evolution, they can represent elaborate methods for dealing with inconsistent data, e.g. the ambiguity between genuine trichotomies and
unresolved dichotomies (Fig. 18). More importantly, tree construction is
regarded as essential in the testing of competing theories of speciation,
because trees alone are composed of species and speciation events. However,
only the cladogram can be converted into a classification, because it does not
assume the presence of ancestors.

Classification construction

In the second aspect of the cladistic method, it is argued that the
cladogram is directly converted into a classification. The classification is
based on the reconstructed phylogeny. Variation has arisen, however, in how
best to do this in the light of fossils, since the latter's inclusion or exclusion
greatly determines ways of maximising information content in a classification.

Hennig originally argued that in converting the cladogram into a classification, the taxa rank must be determined by the relative times since the
component species of the taxa diverged from a common ancestral population. Indeed, Hennig regarded this as the *fundamental construction* in which
the categorical rank of a taxon is automatically given by the relative age of
the stem species. Such measures were a direct result of Hennigs's definition
of strict mónphyly (and implied that sister group taxa must be coordinate –
of the same absolute rank, and his desire to include fossils in a classification).
Referring back to Fig. 9 the classification would be as follows:

Fig. 15 The outgroup comparison.[35]

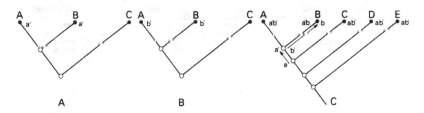

I   *Taxon*   Z

II   *Taxon*   (X–Y)

    Taxon Y

    Taxon X

Hennig's system of ranking was not widely accepted since it led to the piling up of categories. Present day phylogenetic cladists see three flaws in Hennig's subordination scheme.

(a) Because each branch of a phylogenetic diagram necessitates sub-ordination, it cannot accommodate the known taxonomic diversity of most groups of organisms. Such a classification would require the creation of new categorical ranks and the creation of new names for taxa at each of these ranks.[41] McKenna's[42] mammalian classification reflects these problems.

(b) The problem of fossil classifications if coordinate sister taxa[43] are to have the same rank in a subordinated phylogenetic system. This involves the problem of incorporating fossil taxa of low rank as sister

Fig. 16 Synapomorphy as a test of cladistic hypotheses.[36] Black represents the derived condition; black boxes – autapomorphies, black bars – synapomorphies. (A) In these cladograms a' is a postulated synapomorphy, while a is the primitive condition. The first of these cladograms, (a), is regarded as the best of the three because in it the sets are best defined by the synapomorphy. (B) With six postulated synapomorphies (a' – f'), the best cladogram is (d), because it minimises convergences to two; (f) has six convergences to (e)'s four.

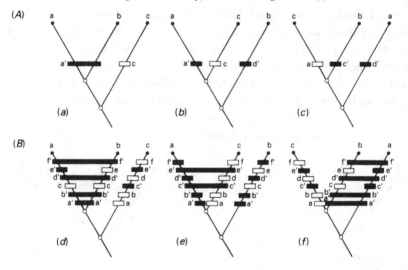

groups (with extant members) of high rank. All too often the rank of the fossil taxa has to be elevated.

(c) If branching within phylogenetic classification is determined by the branching sequence of a cladogram, then new discoveries lead to major rank alterations. Here Hennig's solution was to replace categorical names with a numbering system, thereby constructing a non-Linnaean classification.

Of course, Simpson's phylogenetic modification had anticipated such problems, but phylogenetic cladists had to find alternative solutions because of their adherence to the one-to-one relationship between cladogram and classification. They appropriated several conventions for the exact expression of relationships between taxa in a classification. First, phyletic sequencing,[44] whereby a series of fossil taxa of the same rank within a taxon of higher rank are arranged so that each taxon is the sister group of all those succeeding it. This removes the problem of proliferating categorical and taxa names in the recognition of fossil taxa with recent taxa. Second, the use of the 'plesion' concept.[45] In addition to the normal category names from species upwards, all fossils are denoted 'plesion' – that is, a name of variable rank. This avoids the necessity of a fossil species being the sole representative of a monotypic genus, family or order (which violates the rule that sister taxa must be co-ordinate).

It is clear that phylogenetic cladists aim for the direct and accurate representation of genealogical information in a classification (see Fig. 19). That fossil studies represent a thorn in the side to this enterprise is evident. Not

Fig. 17 A comparison of cladograms and trees.[38] (a) Cladogram: species are depicted as dots; lines connect dots to indicate nested patterns of synapomorphy. (b) Phylogenetic tree: as consistent with the cladogram in (a). Dots represent species, lines connect the species in terms of ancestor–descendant relationships. X – hypothetical ancestor. (c) Phylogenetic tree – species are represented by solid lines indicating known temporal distributions. Dashed lines depict hypothetical lines of descent. (d) Original cladogram redrawn using the notation employed in (c).

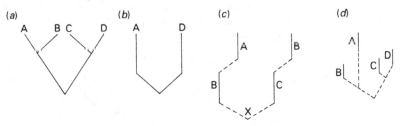

surprisingly, since Hennig, fossils have come to be ignored in both cladistic analysis and classification construction.

The Hennigian view and theoretical information

The problems confronting Hennig are rather different to those that we encountered with Mayr. Hennig adovcated a one-to-one correspondence between the phylogenetic tree (cladogram) and classification, and took great pains in setting up explicit methods which were consistent with his aim of storing as much theoretical information as possible in classifications. What is at issue here is whether it is worth pursuing the ideal of a classification storing as much information as possible.

In arguing for the maximisation of information content in classifications, Hennig is making two distinctive claims. First, his reconstructed phylogenies represent the particulars of evolutionary theory, and in this respect are a representation of a particular theory. Second, classifications are directly derived from the reconstructed phylogeny (the cladogram). These two claims are represented in the distinction between cladistic analysis and cladistic classifications. These are logically separate matters since the former focuses on methods of phylogenetic reconstruction, while the latter concerns the theory of classification.

It is important to realise that even if the methods of cladistic analysis do not give an accurate representation of phylogeny, this would not be inconsistent with the claim of having maximal information content in a classification. Of course, if it is argued over and above this, that phylogenetic cladists must uncover the true phylogeny, then they should present a representation of evolution that is the most likely one.

The problems surrounding phylogenetic cladistics concern what can and cannot be represented in cladograms and classifications, as well as the complexity of such rules of representation in determining what is represented. I will therefore concentrate on:

Fig.18 Trichotomies and unresolved dichotomies.[40] (a) Cladogram representing two dichotomies. (b) Cladogram representing a pair of unresolved dichotomies, a genuine trichotomy, common ancestry and reticulate evolution. (c) A cladogram which expresses doubts over relationships. In time (b) and (c) may be resolved either as a genuine trichotomy or a pair of dichotomies. The latter is the only permissible alternative for the cladist.

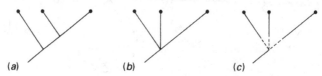

(a)   (b)   (c)

(i) whether cladistic methods for inferring phylogenies are one way of ascertaining phylogeny, and even the only way of ascertaining phylogeny or knowledge of evolutionary phenomena;

(ii) whether the cladistic approach to classification allows for the accurate representation of the information derived from cladistic analysis.

I will argue that cladistic classifications (Hennigian and phylogenetic) cannot represent all aspects of phylogeny, but as it stands, this is not necessarily a fatal flaw. It is a more important matter though, whether Hennig's goals are sensible. Does cladistic analysis give a sensible picture of evolution? Is it sensible to convey as much information as possible in the manner that the phylogenetic cladists do? Is it sensible for a classification to convey these aspects of phylogeny? One unfortunate corollary of the Hennigian system is that the incorporation of fossils is problematic, and the question of whether the aims are sensible is very pertinent in the light of this anti-fossil attitude. This will be examined in detail.

One final point should be made; all references to cladists in this section refer only to Hennig and latter-day phylogenetic cladists. Since the

Fig. 19 (*a*) A cladistic hypothesis for some major groups of vertebrates. Eu – Eutheria; Me – Metatheria; Mo – Monotremata; Av – Aves; Cro – Crocodilia; Am – Amphibia. (*b*) The corresponding classification.[46]

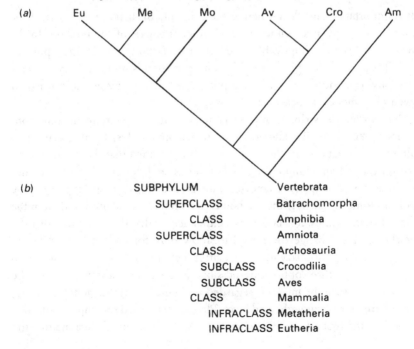

phylogenetic cladists are heavily indebted to Hennig, they, too, are susceptible to similar criticisms.

The representation of phylogeny in cladistic analysis

This involves various phenomena which cladists believe to be unknowable, as emphasised in their methodology for representing the model of evolutionary change. Importance will also be attached to whether the model of process is realistic, and whether it can be represented adequately. It will become evident that the rules permit only an unrealistic model of evolutionary process to be represented.

As we have seen, phylogenetic cladists incorporate an Additive Model of evolution which stipulates that evolution is divergent. But it is important to be clear over the distinction between model and method, since the two are easily confused. This distinction is highlighted in the principle of dichotomy.

The principle of dichotomy reflects the assumption that speciation is dichotomous, and Hennig is quite clear that a dichotomous differentiation of the phylogenetic tree ' . . . is primarily no more than a methodological principle.'[47] Later on, however, he appeals to empirical considerations;

> A priori it is very probable that a stem species actually disintegrates into several daughter species at once, but here phylogenetic systematics is up against the limits of the solubility of its problems.[48]

It is important to be clear whether the principle of dichotomy is part of the model (i.e. a given assumption) or whether it is part of the method. In the former, it becomes impossible to obtain a trichotomy, so Hennig's point is that it is a methodological rule. This interpretation is reinforced by present-day phylogenetic cladists who do not regard dichotomy as the normal or even only mode of speciation.

Yet, within the writings of phylogenetic cladists, there remains some confusion over the model and method. To exemplify this, I will discuss two alternative situations. In the first case, if it is assumed that the model allows for splitting of any number, i.e. *n*-chotomous, and this gives a tree structure without closed loops (i.e. ignoring introgression[49]), then speciation can be polychotomous. Now, if the method is to search for dichotomies, then in the face of trichotomies, it is possible to either accept that this is the way that the world is, or to change the model to allow only for dichotomies. Alternatively, if the evolutionary model specifies dichotomies, and we are left with polychotomies, three options are possible: (i) the polychotomies need not be specified, and yet the model is regarded as adequate, (ii) the model is wrong, (iii) there is not enough evidence. Phylogenetic cladists imply that they adhere to the first alternative, when in reality they implicitly assume that the principle of dichotomy is part of the model. This is problematic. In the first place, even if there was good evidence of multiple speciation, then the principle of dichotomy (in the context of a methodological rule) would be redundant since the theory of evolution was being misrepresented. In other words, if polychotomies are part of the model, they still are not accepted. Cladists do not regard this as a problem because they argue, in addition, that our inability to distinguish dichotomous speciation from more complex sorts of speciation and our inability to represent unequivocally more complex sorts of speciation in cladograms and classifications, justifies the retention of the principle of dichotomy. Thus, even if multiple speciation does occur in nature, a cladist never feels justified in concluding that a speciation event is trichotomous because he is never justified in dismissing the possibility that an unresolved trichotomy is really a pair of unresolved dichotomies.

Clearly, this line of argument is misleading because it presupposes that we can never come to know a trichotomous speciation, but that we can be sure of a dichotomous speciation. No doubt the infrequency of trichotomous speciation events implies, to the cladist, that we are more certain in our discovery of dichotomies. But the frequency of a particular phenomenon should not colour our method of discovery of such phenomena. It is perfectly plausible that we could find a genuine trichotomy rather than an unresolved trichotomy, just as it is perfectly plausible that a dichotomy could represent a single lineage mistakenly divided into two. If a systematist can be reasonably sure that he has discovered a genuine trichotomy, then there is no reason that he cannot be sure that he has found an unresolved trichotomy – even using cladistic methods. It appears, then, that the principle of dichotomy is taken to be part of the model, despite claims to the contrary. It is also possible that types of evidence other than phylogenetic could be used for investigating trichotomies, such as Platnick & Nelson's[50] suggestion of data from historical biogeography. If, in both sources of data, there is a congruency – in the form of a trichotomy – then should we not take this as evidence of a trichotomy?

The failure of the phylogenetic cladist adequately to come to terms with the problem of trichotomous speciation means that there is an ambiguity in what he attempts to represent. If genuine trichotomies were included as part of the model, then there would be ambiguity in representing either a genuine trichotomy or an unresolved dichotomy. In arguing against trichotomies as real phenomena, cladists are predisposing themselves to a particular view of evolution; thus, if speciation is regarded as the gradual splitting and divergence of large segments of a species, as Hennig[51] maintains, then multiple speciation looks less likely than if it is viewed as a process in which small, peripheral populations become isolated and develop into separate species as Wiley[52] supposes. The phylogenetic cladists' lack of clarity over model and method is all too clear, as well as the dubiousness of specifying dichotomy in the model.

Phylogenetic cladists document several types of evolutionary phenomena that cannot be easily represented through cladistic techniques; ancestor–descendant relations, phyletic evolution and speciation, and deviation in one of two daughter species after speciation (the deviation rule).[53] Justification for cladistic methods of inferring phylogenies is largely methodological which, in some cases, is presumed to be empirically based. There is a danger, here, in interpreting methodological principles as empirical beliefs. Out of the phenomena just listed, ancestor–descendant relations do exist, multiple speciation is likely, and speciation without deviation is unlikely (although the unique derived character may be all but undetectable). All too often, the cladistic justification for method appears to be much greater than it actually is, giving rise to what Hull[54] has called 'methodological shoring-up'. Strictly speaking, an empirical claim asserts a fact; in the case of ancestor–descendant relations, if it is asserted that they never occur, then this is an empirical claim. If it is asserted that we can never distinguish between recency of common ancestry and ancestral–descendant relations, then this is a methodological claim. Finally, the claim that ancestral–descendant relations, if they do occur and can be discovered, cannot be represented unambiguously in cladograms and classifications, concerns the limits of such modes of representation.

I do not want to discuss methodology in detail here (see Chapter 6) but only to emphasise the distinction between the particulars of evolutionary history (as reported empirically) and the methods of their representation. If there is little in the way of a correct empirical basis in cladistics, but instead a greater emphasis on methodology, then it is difficult to see what cladistic models represent, with nothing but a plethora of rules and regulations.

(a)   *Ancestor–descendant relations.*

It is universally acknowledged that certain species which once existed no longer exist and that some of these species gave rise to Recent species. Cladists, however, raise two objections to ancestor–descendent relations:

(i) ' . . . ancestral species cannot be identified as such in the fossil record';[55]

(ii) ' . . . they are also inexpressable in classifications'.[56]

These objections are clearly methodological and emphasise the tension between difficulties in discerning ancestor–descendant relations and problems in representing them. There is, of course, no empirical justification for their treatment of ancestor–descendant relations.

A second distinction of equal importance is between the recognition of extinct species as species and deciding which species gave rise to which. Engelmann & Wiley[57] recognise extinct species as species but maintain that the ancestor–descendant relation is unknowable. In this case there seems little justification for knowing only the former, and not the latter. Those cladists who argue that neither is knowable are spared this dilemma but are again faced with the problem of what to do with fossils. What is of importance here is the recognition that these methodological claims have no basis in observation, and reflect instead an underlying prejudice against fossils.

There is a further problem concerning the accuracy of representation of such relationships. In Fig. 20(a), if species B were directly ancestral to C, these taxa would be expected to share some derived features not found in A. But all this would establish, for the cladist, is that B and C are more closely related to each other than to A; B might be either an ancestor of C or a sidebranch, i.e. a sister species of the ancestor of C (Fig. 20 (b)). Sidebranch status is confirmed only if B has derived features that are not found in C (Fig. 20 (c)). For the cladist, absence of derived features does not imply that B and C have ancestral–descendant status – only that it is an unresolved sister-group. Thus methodological rules here favour one phenomenon (recency of common ancestry) over another (ancestor–descendant relations) with little empirical basis.

(b)   Phyletic speciation.

The choice of one evolutionary phenomenon over another is most clearly represented in the phylogenetic cladists outright rejection of phyletic speciation. Simpson[59] argued that lineages sometimes change sufficiently through the course of their development so that later stages could be considered distinct species even though no splitting took place. A good example can be found in the lineage of fossil Silurian brachiopods of the genus *Eocoelia*.[60]

Objections to this view are couched in operational terms: it is not only impossible to divide a continuously evolving lineage into distinct species in a non-arbitrary manner, but also lineages cannot be established in the first place because there is no way of knowing whether an organism is part of a lineage. There is no clear cut way to recognise species through time in the manner required by evolutionary systematists, since any chopping up of a lineage in time would be arbitrary:

> To attempt to divide a species between speciation events would indeed
> be arbitrary; we could call an individual person by one name at age 10
> and a different name at age 30. Dividing species at their branching
> points, however, becomes not only non-arbitrary but necessary: we
> would not call a son by the same name as his father.[61]

Leaving aside the validity of such an analogy, the point being made is that
new species can only be recognised at speciation events where splitting
occurs. This is clearly a methodological claim since the cladists are stipulating
what we can and cannot represent – we can only recognise new species at
splitting events. Such a view is consistent with the principle of dichotomy.

Fig. 20 Ancestor–descendant relations: 1 = origin of derived features shared by
B and C; 2 = origin of features confined to B.[58]

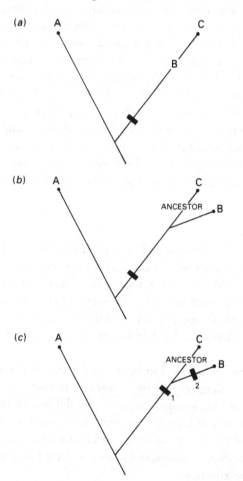

Support for this view of speciation has not all been empirical. Hull supports Hennig's original formulation[62] of speciation, in which the ancestral species is split into two *new* species, not because of any methodological arguments, but because species are individuals[63] – integrated gene pools, and not classes of similar organisms. Indeed, Hennig's discussion is regarded as 'consistent with the most theoretically appropriate definition of species.'[64]

This question of the ontological status of species (as classes or individuals) has been hotly debated in recent years.[65] Originally species were interpreted as natural kinds, universals, secondary substances, spatiotemporally unrestricted classes and the like. But Ghiselin and Hull have argued that species are 'spatiotemporally localised, cohesive and continuous entities'[66]. Organisms are not just members of the class of species to which they belong, but represent parts that go to make up the constituent whole, just as cells, tissues and organs are parts and not members of individual human beings. Hence,

A continuously evolving lineage should no more be divided into distinct species than an organism undergoing ontogenetic development should be divided into distinct organisms.[67]

In this quote, and in the one above from Platnick, there is a prevalent misconception that it is valid to compare species with organisms by analogy. There are several difficulties with this view: first, a problem in arguing over this part–whole and membership relation is that linguistically there is no necessary distinction over parts and individuals. It is possible to make classes out of the species as an individual since a class can be defined with respect to a part of an individual. Thus, the human body can be seen as an individual with respect to its constituent parts, such as cells, tissues and organs, and yet these same parts can be recognised as members of a class. In effect, the part/individual distinction can be shuttled around, with no possibility of 'anchoring' at any level. Secondly, Hull's argument relies on a comparison of organisms with individuals, followed by a comparison of species with organisms. The justification for this comparison is weak since it only relies on the levels at which evolutionary processes are said to operate. In moving from the level of the organism to the level of the species there are no clear-cut boundaries; such a 'haziness' of levels is seen to substantiate this 'how like organisms argument', and the ensuing conclusion that organisms and species are entities of the same logical type.

This is misguided: a comparison of species and organisms does not do justice to the range of types of individual that can be specified. Individuals can either be abstract or concrete objects. To understand fully the implications of this distinction it is necessary to mention Frege's original characterisation of objects. For Frege, objects are the referents of proper names such that the truth conditions of sentences containing proper names are to be explained in terms of the relation between proper names and the objects for which they stand. The subsequent distinction made between abstract and concrete objects is described by Dummett as follows:

> It is . . . a sufficient condition for something to be a concrete object that it should affect our senses, and can, therefore, be referred to in terms of its sensory impact. As a necessary and sufficient condition the object be perceptible to some conceiveable sensory faculty. . . [or] be detected by some instrument or apparatus; and this amounts to no more than saying that the object is one which can be the cause of change.
>
> More generally, a concrete object can take part in causal interaction; an abstract object can neither be the cause nor the subject of change.[68]

Entities such as shapes, electrons, are concrete objects in the sense of being real, but at the same time, species and electrons could be abstract objects by virtue of their role in theories. In this respect, abstract objects are pure sets.

Returning to Hull's argument, it is clear that he follows the following argumentative chain:

(i) objects are the referents of proper names;

(ii) species are the referents of proper names;

(iii) *ipso facto*, species are individuals.

The fallacy of comparing species to organisms is now clear; organisms, like chairs are concrete entities, and it follows that since organisms are concrete entities, then species must also be concrete entities. Consequently, there is no theoretical status for the species as an individual because Hull assumes that all individuals are concrete. Yet, much of his argumentation incorporates a theoretical notion of species.

Arguments over the ontology of species, especially when based on a comparison of species with organisms, do not resolve whether or not Hennig is justified in rejecting phyletic speciation. This is an empirical issue and must be treated accordingly. Nor do arguments directed at our inability to recognise species between speciation events refute the claim that phyletic speciation does not exist. Both phylogenetic cladists and Simpsonian systematists acknowledge the difficulty of recognising species through time; but just because it is difficult to detect a particular phenomena does not imply that it does not exist.

In effect, the question of phyletic speciation boils down to a question of method, and representation. Cladists stipulate that new species can only be recognised when splitting occurs. In non-saltative speciation, temporal continuity is maintained, but the cohesiveness of the lineage is disrupted.

Further to this, whenever splitting takes place, the ancestral species must be considered extinct. According to Bonde, Hennig's species concept is the

> only logical extension in time of the concept of the integrated gene pool. At speciation this gene pool is disintegrated and two (or more) new sister species originate, while the original species becomes extinct.[69]

Wiley agrees that

> in most cases the methodological necessity of postulating extinction of ancestral species in phylogeny reconstruction as advocated by Hennig (1966) is biologically (as well as methodologically) sound.[70]

However, Wiley prefers to use an evolutionary species concept in which an ancestral species can survive a split, if it can lose

> one or more constituent populations without losing its historical identity or tendencies.[71]

Under Hennig's view there will be more species, while for Wiley there will be fewer.

If there is massive disintegration of the gene pool at speciation, then Hennig is on safe ground. But there is evidence from Mayr[72] and Carson[73] that speciation occurs through the isolation of a small peripheral population, with little, if any, disruption of the parental gene pool. Empirical evidence is again at odds with Hennig's methodological proviso that all ancestral species become extinct at speciation. Of course, there may be some instances when Hennig's views are correct, but this does not justify the choice of one mode of representation of a particular set of evolutionary facts over another. The method does not fit the empirical facts.

(c)   *The deviation rule.*
        According to Hennig

> When a species splits, one of the two daughter species tends to deviate more strongly from the other from the common stem species.[74]

Brundin[75] sees this as one of the fundamental principles of life, while Schlee[76] argues that it is inessential to phylogenetic cladistics. Nelson[77] agrees with Schlee but adds that it can function as a methodological principle.

The first thing to note about the deviation rule is that it is curiously at odds with Hennig's convention that even if an ecologically isolated population gains genetic isolation from its parent stock, both should be regarded as new species.[78] Secondly, the deviation rule is plausible in nature in that a recently split population may diverge more greatly than the parent population, in a rapidly changing environment. More important, though, is the methodological form of the rule; it is irrelevant whether this rule is true empirically because cladistic analysis requires at least one unique derived character in one of the groups. If no character can be found, then cladistic methods will not be able to distinguish between the groups. As a methodological rule this is on strong ground since recognition criteria of species are often restricted to morphology, biochemistry and behaviour. On this basis it is often difficult to distinguish between sibling species, and it is plausible that if two species are known to be reproductively isolated, there still may be no character which distinguishes between the two. In this respect Hennigian methods share the same problems as other methods which attempt to recognise species. There is always a danger that two groups may not be separate species (splitting groups) or that one species may actually be two.

In conclusion to this discussion of phylogenetic cladistic methods for representing phylogenetic information, it should be clear that in the examples mentioned there is not much empirical justification for the methodological rules. The two are not always closely connected. While this in itself is not necessarily a grave error, the consequences of such attitudes can be, since it is not apparent that such rules are sensible. This is all the more pertinent since phylogenetic cladists emphasise clarity and unambiguity of model and methods, yet the ambiguity still remains.

The relationship between cladogram and classification

It cannot be denied that cladists have had problems in representing certain features of evolutionary development in the Linnaean hierarchy. Cladists regard this as a limitation of the Linnaean hierarchy, and instead of refusing to represent what a particular system of representation has difficulty in representing, some have suggested that a better alternative might be to improve or abandon that system of representation. Favouring the latter course of action, both Hennig and Griffiths[79] have advocated the abandonment of the Linnaean hierarchy, with a replacement based on a numerical or unclassified hierarchy of taxa. Less radical cladists[80] have attempted to formulate various improvements based largely on the problem of incorporating fossil organisms into a classification. It remains, however, a moot point for the cladist whether or not the representation of the particulars of phylogeny in a classification by direct correspondence is sensible.

The advocacy of alternative modes of representation is nothing new in taxonomy (e.g. non-hierarchical systems and 'quinarian' approaches to classification[81]), but it is generally acknowledged that the Linnaean hierarchy is the best kind of classificatory system used at the supraspecific level. This is not only a matter of historical precedent, but also because the Linnaean hierarchy gives convenient diagnosis and information retrieval. Most important, though, is the observation that the morphological dissimilarities between taxa of specific rank often exhibit significant hierarchical structure.[82] Calls for the modification of the Linnaean system should therefore be treated with close scrutiny.

The claims made by the phylogenetic cladists are suspiciously self-preserving. Is it really a failing of the Linnaean hierarchy that it cannot represent the particulars of cladistic analysis? It could be said, equally, that the methods of cladistic analysis should be altered to accommodate the Linnaean hierarchy, and that any other course is not sensible. The results of the discussion on cladistic analysis certainly undermine the pursuit of inferring sister-group relations from biological classifications to the extent that the increase in complexity and asymmetry of the resulting classification is not warranted. Yet cladists are very reluctant to confront this problem because of their fundamental assumption that the cladogram and resulting classification have a similar logical structure. There must be a direct relation between the two, with no loss of information content;

> Cladograms and Linnaean classification schemes share a similar logical structure: they consist of sets within sets. In cladograms the nested sets of taxa reflect nested sets of synapomorphies; in Linnaean classifications, the nested sets of taxa are the primary information, for the classification itself could have been based on a branching diagram derived from one of a variety of approaches.[83]

Only the particulars of cladistic analysis can be accurately reproduced in a classification. For the cladist, then, there is little reason to question the fundamentals of cladistic analysis because they are already determined by the logical structure of the Linnaean hierarchy.

But such a view is an entire misconception: *cladograms and classifications do not have a similar logical structure*. The roots of this misconception can be traced back to Hennig's own discussion of hierarchical systems, which is based on the Woodger–Gregg model.[84] Using Woodger's definition of a hierarchic type of system, Hennig[85] argues that it is an adequate representation for the phylogenetic relationship between species. In Fig. 21, the structure of the phylogenetic relationship corresponds to the Woodger–Gregg definition of a hierarchy in the following way:

> the elements $(x_0, x_1, \ldots, x_9)$ composing the ordered quantity in the system are paired by relations that extend in only one direction.[87]

This would be permissable if the Woodger–Gregg model did give an adequate account of the Linnaean hierarchy. But it does not. Consequently, since Hennig, all cladists have been in the uneasy position of arguing that the Linnaean hierarchy is tailor made for their purposes, while at the same time attempting to modify the Linnaean hierarchy because it leads to problems of representation.

Problems in the Woodger–Gregg model

The model states that a set of individual organisms is successively partitioned, the subsets being taxa, and the number of steps taken to reach a given taxa being the measure of the rank of the taxon. The set of taxa of a given rank constitutes a category.[88] In this formulation taxa are regarded as sets of individual organisms.

Many biologists have supposed that the Linnaean hierarchy raises special logical and philosophical problems, most notably in the practice of monotypic classification. Gregg[89] showed that the existence of monotypic taxa leads to contradictions within the model. It is impossible to distinguish, on the basis of rank, between taxa which have the same individual organisms as members, e.g. membership of the monotypic taxon Ginkgoales, of ordinal rank, would be the same as of the taxon *Ginkgo biloba* of specific rank, so that the two taxa would have to be regarded as identical. In monotypic classifi-

Fig. 21 [86] (*a*) Structure of the phylogenetic relationship; (*b*) structure of the relationship among the elements of a hierarchy based on Gregg's model. Note in (*b*) that the point of one arrow can lie in each element of the hierarchy, whereas several arrows may arise from it.

(*a*)

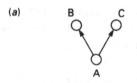

(*b*)

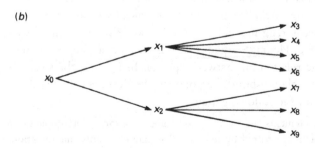

cations, two taxa with exactly the same members are classed at different category levels, so that every case of monotypic classification will generate an overlap between two categories which are intended to be mutually exclusive. Buck & Hull[90] traced this failure to Gregg's employment of set theoretic procedures for specifying the meaning of taxonomic names, and his criterion of identity for taxa, whereby two taxa are identical if they have all and only all the members. They argued that it was fallacious to treat taxa as sets which are defined extensionally – the extension of a class name simply being the set of members of the class. Instead, they recommended an intensional definition such that that the intension is the set of properties which qualify an entity for membership. Such definitions specify the sets of properties which supply the qualification for membership.

Jardine,[91] however, has argued forcibly that 'Gregg's paradox' is nothing more than an inadequacy of the model, and that the construction of an adequate set-theoretical model for the taxonomic hierarchy presents no special difficulties. The problem in the accounts of Gregg and Buck & Hull stem from their attempts to clarify simultaneously both the logical relations between taxa in a hierarchy and the relationship between taxa and individuals. This is not necessary (as mentioned in Chapter 1) because the relation between taxa individuals is not based on set-membership, despite the fact that we speak of individual organisms as 'belonging to' taxa. (A more correct terminology would be that *Homo sapiens* is a member of the extension of the taxon Mammalia.) Taxa, like other empirical class concepts, cannot be defined in terms of any set of properties – if they must be defined, it should be extensionally, and such definitions do not contribute anything to a logical account of identification, since such an account must show how individuals are assigned to taxa on the basis of their observed characters. So, to define taxa extensionally in terms of individual organisms hinders describing the logical structure of taxonomic hierarchies. Jardine concludes by saying that the source of much of the confusion over the nature of taxonomic hierarchies is a failure to realise that it is possible to give an account of the logical structure of taxonomic hierarchies without an account of the way in which individual organisms are diagnosed into taxa.

Thus, monotypic taxa can be included in a logical account of the taxonomic hierarchy (Fig. 22). The points in the diagram represent the extension of taxa, and the levels (0, 1, 2, . . . $N-1$, N) at which points occur indicate the rank of the corresponding taxa. Two points are connected by a line if, and only if, the extension represented by a point at the lower level is included in the extension represented by the point at the higher level. The extension of one taxon is included in the extension of another taxon if, and only if, every basic taxon which is a member of the extension of the first taxon is also a member of the extension of the second taxaon. Thus a taxon is specified by its rank and its extension.

Problems in ranking

Because of the difficulties in the Woodger–Gregg model over the status of monotypic taxa, if phylogenetic cladists are to argue for a direct correspondence between cladogram and classification, then the cladogram must be symmetrical (Fig. 23(*a*)), giving the minimal number of ranks per number of terminal taxa. However, not all cladograms are symmetrical, and in the case of fossil species, the cladistic criterion of ranking leads to an asymmetrical cladogram (Fig. 23(*b*)) with a maximum number of ranks. This is a direct consequence of the cladistic criterion for ranking which specifies that sister-groups must always be given co-ordinate rank.

In the initial formulation of this criterion Hennig argued that the rank of the taxa be determined by the relative times since their component populations diverged from a common ancestral population. In the case of fossil species, there is a problem; early groups, even if they admittedly became extinct without leaving descendants, would have to be recognised as separate phyla, equivalent to highly diversified, persistent groups. If it could be shown that a species splits off in the Precambrian, but gave rise to no other species, it nevertheless would have to be classed as a phylum. The resulting classification would be exceedingly monotypic. With such a proliferation of monobasic group names, the cladist is open to a charge of conceptual redun-

Fig. 22 A generalised representation of a taxonomic hierarchy.[92]

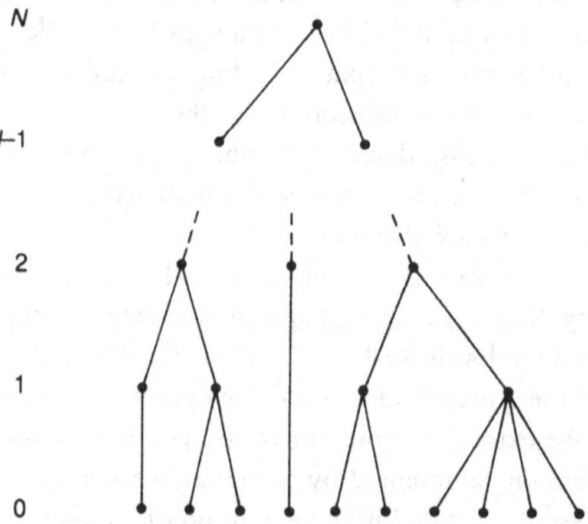

dancy.[93] This is clearly a contradictory position for phylogenetic cladists because they adopt a model of classification that is incapable of accommodating monotypic taxa, and yet their cladograms, in some instances, require a classification that is monotypic in the extreme.

Now it is true that the Linnaean hierarchy assumes that species occupy a fixed level in relation to taxa of other ranks, and that this is incompatible with a model of phylogeny which is derived from truncated hierarchies in which the end-points (representing the extinction of terminal species) are not all equidistant from the beginning of the phylogeny (representing the origin of the ancestral species). But this incompatability only becomes a problem if it is assumed that sister groups must always be of coordinate rank, with the resultant piling up of categories. The phylogenetic cladist is mistaken in his belief that classifications can be used to express evolutionary branching sequences alone.

It should be clear that phylogenetic cladistics is not impossible in principle with a strict Linnaean hierarchy (that excludes monotypic taxa and is based on fixed ranking) and various stringent methodological rules. But is it really necessary to 'hijack' the hierarchy in this manner? Such a strategy is surely misguided in the light of the fossil record, and its subsequent rejection as a means of avoiding the piling up of categories.

The problem of fossils

Arguments against the use of fossils in classification construction have been presented both by phylogenetic and transformed cladists. Reference will be made to the latter's arguments since they are of direct relevance to the former. The rejection of fossils by cladists occurs at two levels; in cladistic analysis and classification construction. Both aspects will be examined here.

As mentioned in Chapter 2, attitudes to fossils are couched in terms of the

Fig. 23 The two extreme forms of cladogram shown for eight hypothetical terminal taxa: (a) symmetric; (b) asymmetric.

(a)   (b)

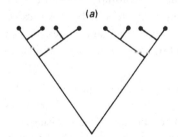
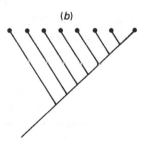

paleontological and neontological approaches. For any cladist, this implies that paleontology is consistent with the construction of trees, and is justified through evolutionary theory. In contrast, neontology is concerned with cladogram construction and the discovery of groups.[94] Crucial to this distinction is the definition of relationship employed (Fig. 10). Patterson[95] describes two types of relationship for organisms of different species, or of the same asexual species. In evolutionary terms, there are relationships of common ancestry (Fig. 10(a)) and by direct descent (Fig. 10(b)). In addition (a) specifies the non-evolutionary relationship used by transformed cladists.[96] For present purposes, however, the fundamental difference in attitude to fossils is expressed in the distinction between (a) and (b), irrespective of whether the former is evolutionary or non-evolutionary.

For Patterson, evolutionary relationships can only be obtained by character distributions in present-day organisms. In an explicit statement on the matter, he concludes that

> instances of fossils overturning theories of relationship based on Recent organisms are very rare, and may be nonexistent. It follows that the widespread belief that fossils are the only, or best, means of determining evolutionary relationships is a myth . . . It seems to have followed, as an unquestioned corollary, from an acceptance of evolution.[97]

Systematic arguments are backed by an historical analysis to show that paleontological data has been more of a hindrance than a help in determining evolutionary relationships. I will first say a little on the historical arguments.

In attempting to erect a standard by which to judge the influence of fossils, Patterson examines historical case studies of teleostean phylogeny[98] and tetrapod evolution,[99] concluding that

> since Darwin we have been too eager to think in terms of origins, and to view problems such as how vertebrates got on land as the province of the paleontologist, soluble through discovery of fossils intermediate between extant groups.[100]

He goes on to argue that pre-Darwinians gave paleontology a subservient role to neontology (in the sense of examining relationships between Recent organisms), and that attempts to classify using characters such as fish scales, which could be investigated in the fossil record, were mainly ignored. But with the advent of evolutionary theory a subtle shift occurred. As Ghiselin sees it,

> The change Darwin made exemplifies a basic shift in attitude. Instead of finding patterns in nature and deciding that because of their conspicuousness they seem important, we discover the underlying mechanisms that impose order on natural phenomena, whether we see that order or

not, and then derive the structure of our classificatory system from this understanding. The difference, then, lies with the decision as to what is important.[101]

Before Darwin, the key to pattern as manifest in abstract relationships, was homology, but this became transformed into an examination of the real descent of one group into another, as evidenced by the stratigraphic sequence. Under this research programme increasing attention was paid to fossils, as couched in terms of ancestor–descendant relationships, and to the identification of actual ancestors as the goal of paleontology. The paleontological method was seen as the postulation of transformations in stratigraphic series of fossils. Thus, what concerned the pre-Darwinian paleontologists was how fossils were related to Recent species, not how Recent species were related to fossils.

Patterson, however, fails to point out two details in this historical scenario: first, the major reason why the pre-Darwinian systematists were uninfluenced by the fossil record was due to its very poor quality. Second, the notion of modification by descent emphasises an important factor – time, so that the fossil record documented the history of life.

More importantly, Patterson claims that the search for ancestors has been unsuccessful; one putative ancestor after another has been rejected without any greater knowledge of the 'real' sequence of ancestors being achieved. In contrast, all decisive advances in teleost phylogeny have arisen through neontological analysis. A similar claim is made for the fish–amphibia transition.[102] Here, Patterson regards the Dipnoans, and not the Rhipidistia, as more closely related to tetrapods. Yet, this conclusion is based on fossil material – the Gogo lungfish *Griphognathus*, from the Devonian – and it is only because of this specimen that the 'Rhipidistian barrier' was broken. Patterson does not practice what he preaches. Fossils can and do influence evolutionary relationships based on Recent organisms.

Patterson is at pains to back up the contention that advances can only occur through neontology alone, as well as the methodological claim that ancestor–descendant relationships cannot be adequately represented (and that fossils are therefore useless). No historical analysis based on case studies can prove this, especially in the light of Patterson's strategy for dismissing fossils, in which the correctness of neontological analysis is assumed *a priori* and used as a means for evaluating the contribution of fossils. Now, in either case, conclusions based on neontology or paleontology could be wrong, so that if these two attitudes are to be compared, it must be by an independent standard, and not solely through historical analysis. Patterson's argument amounts to the following: given fossils, we must corroborate through neontology. But neontology does not require corroboration from fossils. Therefore fossils are unnecessary. Such a strategy is misguided on several counts. First, the idea that paleontology is irrelevant to the evidential basis of neontology is wrong, e.g. as with *Griphognathus*. Second, the view that paleontological evidence cannot test their results. This is not only false, but also naive to separate neontology from the fossil record. I shall return to this point later on.

In arguing against the view that

> evolution is a theory about the history of life; evolutionary relation-
> ships are historical relationships; fossils are the only concrete evidence
> of life; therefore fossils must be the arbiters of evolutionary relation-
> ships,[103]

cladists present several reasons, ranging from practical to methodological. For the cladist, lines of descent amongst fossils have to be established and there is just too little evidence to do this by the only solid technique available – namely, cladistic analysis. More specifically, cladistic arguments usually centre on the stratophenetic approach,[104] which

> combines stratigraphic evidence of relative temporal position with
> phenetic clustering and linking to yield an essentially empirical reading
> of phylogeny.[105]

For evolutionary systematists this method is far from fallible, and phylogenies 'based on a dense and continuous fossil record are usually very stable and reliable.'[106]

In response cladists argue that the stratophenetic approach is limited to a very small fraction of organisms, while cladistic analysis can be applied to all groups equally. This is plausible enough, despite the fact that the strato-phenetic approach is specifically designed for fossil sequences. Secondly, cladists argue that ancestors often occur in sequences later in time than their descendants so that primitiveness and apparent ancientness are not neces-sarily correlated. This is true, but while primitiveness and apparent ancientness are not always correlated, more often than not they are, so that this criticism is by no means fatal to the stratophenetic approach. Thirdly, ancestor–descendent relations are unknowable regardless of how dense and continuous the fossil record might be.[107] Despite the fact that ancestor–descendant relations cannot be unequivocally represented in cladograms, this is still a puzzling criticism, given the phylogenetic cladists' interest in conti-nental drift and particular modes of speciation.

For all this, cladists are wrong if they assert that cladistic analysis is the only way to get information from fossils. Morphological series, based on a chain of organisms in which the members adjacent to one another are more similar to one another than to those lying elsewhere on the chain, can be combined with a time scale to give polarity sequences ranging from primitive to derived states. This in itself is a possible alternative to cladistic analysis and fundamental to it is character analysis,[108] which underlies any consideration of phylogeny and which involves homology and morphocline polarity. Crucially, while cladists state such sequences may be fraudulent, or that all the information depends upon assumptions about gradualism, they themselves are not immune from similar criticisms. Phylogenetic cladists have an intense fascination with punctuated equilibria and, ignoring the pros and cons of such a view, if gradualism is a rare occurrence, then cladistic premises concerning limited parallelisms and lack of reversals in characters is just as dubious as gradualist assumptions concerning the tempo and mode of evolution. Furthermore, cladograms are sensitive to theories of speciation.

To summarise, the phylogenetic cladists do not have a positive case against the relevance of fossils with respect to cladistic analysis. Of course, it is possible to attack morphological series because they require homology, and if performed in some Recent organisms, such series would be misrepresented if they had come from the fossil record. A cladist would, no doubt, concede the value of reconstructing evolutionary relationships among Recent organisms. There remains, however, a strong case for the phylogenetic cladist to isolate cladistic analysis from an appeal to fossils in which the results of cladistic analysis are used to test interpretations based upon fossils. And the results of paleontology are used to test relationships derived by cladistic methods. Fossil studies and taxonomy are thus separate disciplines which must be kept apart. But why should phylogenetic cladists do this and what is the justification for such a move? Clearly, it is not normal practice to keep two lines of evidence apart like this and there have to be sound reasons for such a strategy. Cladists might say that they are discovering a new technique which needs a certain amount of refinement before the larger questions about evolutionary rates and fossils can be tackled. This is a naive view, though, for even if the larger problems are supposedly ignored, often one cannot but help becoming involved in the big questions.

Alternatively, a cladist could argue that his results may not have a special use for the classification of living organisms. This is a radical idea which presupposes that cladistics and taxonomy are independent disciplines. Unfortunately, this is untenable for several reasons, the most obvious being that this still does not justify the dismissal of fossil evidence when examining a classification based on Recent organisms. Furthermore, do we want a taxonomy which is incapable of dealing with fossils? Obviously not, if we are attempting to derive an objective classification in the sense that the cladists specify. Finally, what makes the phylogenetic cladist think that reconstructing the topology of the phylogenetic tree of living organisms is the aim of taxonomy? There is no justification that this gives the true picture of evolution. Nor is it sensible to reconstruct the topology of trees with little attention paid to the general storage and retrieval of information, whatever the nature of such information. Cladistic analysis is too rigid for general use, and this is even more evident in cladistic classification, where fossils are rejected because they lead to the piling up of categories. If the techniques cannot accommodate the data, there is no justification for rejecting such data. Should not the techniques be reassessed? With their notion of strict monophyly, phylogenetic cladists want to eliminate polyphyly; but at what cost? Much of what phylogenetic cladists want can be obtained using Simpson's technique in which overall resemblance classifications can be adjusted to remove gross polyphyly.

There is one further point which makes the dismissal of fossils even less sensible, and this concerns testability. In discarding the fossil record an essential technique of testing is lost. Without fossils, how are branching sequences testable? With the phylogenetic cladists' insistence on testability, this is indeed strange.

In conclusion, then, the rejection of fossils in cladistic analysis and cladistic classification is not justified. Just because the techniques of analysis and modes of representation cannot accommodate fossils, this in no way justifies their dismissal. Such a strategy is illogical because it also removes an essential technique of testing, while at the same time requiring that trees are unrooted to avoid the question of fossil evidence.

In this discussion of Hennig and latter-day phylogenetic cladists, I have argued that their aims are not sensible because firstly, they cannot represent the true phylogeny in all its aspects, and secondly, they have to 'hijack' the Linnaean hierarchy to uphold the one-to-one correspondence between cladograms and classifications. These cladists fall foul of their own rules of representation and what it is they are attempting to represent. Some of their methodological rules lack empirical content, and do not give unambiguous results. Furthermore, since there are no strict logical relations between cladograms and classifications, classifications cannot be regarded as a short-hand for branching sequences. Thus, in 'hijacking' the hierarchy, phylogenetic cladists assume that their method is the *one* method in taxonomy. But being able to infer sister-group relations from classifications is not worth the increase in complexity and asymmetry of the resulting classifications, nor is it worth the rejection of fossil data and the subsequent loss of information content. Indeed, there seems little point in advocating a strict Linnaean hier-archy which cannot accommodate fossils, especially when their methods cannot sustain it. The failure to accommodate the results of cladistic analysis within the methods of cladisitic classification do not justify claims for the maximisation of information content in classifications.

Furthermore, Hennig's requirements can be performed perfectly adequately using other methods. As we have seen, cladograms can only give minimum distance reconstructions for data, and must depict branching sequences which are rooted.[109] The lengths between the nodes are of no significance, so that a cladogram is a minimally connected graph or tree which is directional. In contrast, trees are non-directed graphs which may or may not be rooted. In advocating cladograms over trees, it is not clear that branching patterns *alone* can give a direct correspondence between cladogram and classification. If we are interested in upholding consistency between the classification of groups of organisms and their phylogenies, then the tree-form or topology of the phylogeny performs this task perfectly adequately, without the incorporation of complex rules that fail to give the standards required.

In addition, equally useful results can be obtained using the Wagner method rather than Hennig's method. If the hierarchic form of a Wagner tree is used directly as a classification, then the classification groups according to synapomorphy. But this cannot give a solution to possible incompatibilities in character states, since the undirected network corresponds to a wide class of trees with the same order.

Finally, many of the evolutionary assumptions that phylogenetic cladists make are gratuitous and false, such as deriving the topology of trees from present day organisms. It would surely be better to make as few assumptions as possible, and then test these against further assumptions. Just stipulating that evolution is conservative is inadequate.

Conclusions

We have seen that to call classifications theories is misguided, and that the positions of Mayr and Hennig do not in practice represent this extreme. They occupy the middle ground. Yet both advocate that not only should the information stored be phylogenetic, but that their respective systems are more phylogenetic than the other. This is an unsatisfactory form of argumentation because it converts an argument based on the 'goodness' of biological classification to one that concentrates on method alone. Thus, classification A is better than classification B because it was obtained using method x, which is known to be better than any other method because it produced classification *A*, which is obviously better than classification *B* because it was obtained by using the method *x*. The danger of circularity is obvious since the method is used to test the result, and not vice versa.

In the case of Hennig, the aims and purposes are not sensible, while the method is too rigid and precludes too many empirical data. In contrast, Mayr's aims are not concordant with his methods, and frequently methodological rules are misinterpreted as explanatory statements. In neither case is it possible to give an accurate representation of the particular theory required. Hennig's strategy is more subtle, however, in that if it can be shown that classifications are abbreviated statements of phylogeny, then it is possible to argue that classifications are theories. But the conventions used by Hennig do not give a direct correspondence between cladogram and classification. Mayr, on the other hand, incorporates explanatory reinforcement to argue that classifications are theories. The rationale behind this is straightforward enough – the greater the amount of information retrieved means the greater the number of extensive inferences that can be made with respect to any classificatory system. But if the method is used to test the result, it is easy to smuggle in assumptions from the method into the finished product. Confusion arises because what may represent a given piece of information stored in a classification and what constitutes an inference based on that information may not be explicit.

At its simplest, the information content of a classification amounts to the claim that the best classification is the one which permits the greatest number of inferences. At a deeper level, though, this implies that classifications serve as indexes for the storage and retrieval of information. A classification is a descriptive arrangement not only for conversing about its included objects, but also for storing and retrieving information about them. But quite what can and cannot be stored represents a major area of tension between Mayr and Hennig. For Mayr, the Linnaean hierarchy is supposedly incapable of expressing the specific kind of information that is unique to trees – information that would include genealogy, morphological-adaptive divergence, genetic similarity and specific claims that would express direct ancestry. In contrast, Hennig argues that it is only possible to recover the topology or structure of the classificatory hierarchy such that the information in a given classification is represented by a hierarchy of taxa, and nothing more. Modern-day phylogenetic cladists concur with this view:

> Linnaean classification is nothing more than a system of names hierarchically arranged.[110]

And

> The hierarchical system conveys a pattern of nested sets, defined by the subordination of subsets at a level or rank within a higher . . . level and second by the sequencing of sets at the same level.[111]

Mayr ignores the point that information content in a classification resides only in the logical structure of hierarchic systems in general. Information is the result of objects being classified into sets, and the information content of a classification is the nested sets within it; nothing more can be retrieved. Farris sums up the situation succinctly, when he states that

> No classification, even a phenetic one, is by itself any more than a collection of organisms. No grouping of organisms, however arrived at, can directly express anything more about the organisms than the membership of the groups. Classifications may possess 'information content' only in the indirect sense that a classification may be used as a reference system.[112]

Mayr is misleading, therefore, in coining the term 'information content' to designate both the expression of set membership and the extensive inferences derived therefrom under the guiding light of a background theory. Classifications have a limited capacity for information storage and retrieval; they cannot depict simultaneously genetic similarity, genealogy and adaptive divergence. Nor is it possible to add information to derive an explanation. The failure of an account based on explanatory essences dictates against this.

In Hennig, the underlying fallacy is more subtle. Hennig argued that hierarchic classifications were completely adequate to indicate phylogeny because he incorporated the requirements of hierarchic classifications into his principles of classification. As we have seen there is a tension between classification and cladogram. The two are not directly interconvertible. Also, with such an emphasis on methodology, Hennig cannot make his biological classifications reflect very much about phylogeny without blocking out other, more scientific functions of classifications. It is one thing to let a variety of considerations influence classification construction, but quite another to formulate a set of principles that permits the unambiguous storage and retrieval of information. While Hennig's methods may recover what is required by cladists, there is a lack of relevancy to such information. Adding more information into a classification does not necessarily say much about the world, but rather more about the method.

PART III

The status of descriptive classifications

4

Phenetics and the descriptive attitude

Introduction

The origins of phenetics can be traced back to two separate sources – firstly the definitive papers of Gilmour,[1] and secondly the availability of computers in the 1950s and the subsequent development of phenetic clustering methods for classification. It has been suggested that the French botanist, Michel Adanson (1827–1906), was the founding father of phenetics,[2] but this is misleading since Adanson did not use the numerical approach in the actual delimiting of genera and families.[3] The numerical methods developed by pheneticists had the virtue of being able to handle quantities of data whose structure could not be grasped by simple inspection, and they aimed to eradicate all the more intuitive methods implicit within evolutionary systematics. With the advent of explicit numerical techniques for inferring phylogenics in the late 1960s, phenetics was no longer synonymous with numerical taxonomy. The latter came to have two areas of investigation: numerical phenetics and numerical evolutionary (cladistic) methods.[4] In this section we will be dealing only with numerical phenetics, the grouping by numerical methods, of taxonomic units into taxa on the basis of their character states. I will not discuss the numerical methods involved in any great detail since our concern is with the steps involved in classifying, and the underlying assumptions, rather than with a strict exposition of numerical methods. Such expositions can be found in Jardine & Sibson's *Mathematical Taxonomy* and Dunn & Everitt's *An Introduction to Mathematical Taxonomy*.

The aim of phenetics is to group the objects to be classified into 'natural' taxa. Such natural taxa are based on Gilmour's definition of naturalness:

> A natural classification is one founded on attributes which have a
> number of other attributes correlated with them, while in an artificial
> classification such correlation is reduced to a minimum.[5]

Thus, the members of a particular natural taxon are said to be mutually more
highly related to one another than they are to non-members. Crucially,
pheneticists define relationship as

> similarity (resemblance) based on a set of phenotypic characteristics of
> the objects or organisms under study,[6]

so that the sole aim of taxonomy is to produce classifications which reflect as
accurately as possible the relative similarities or dissimilarities of populations
without regard for their evolutionary relationships. The aim was seen to
encapsulate the objective approach of phenetics, whereby taxonomy was
purged of all subjective elements. Some of the methods required to achieve
this were, to evolutionary systematists, nothing short of dramatic.

In the first place, pheneticists argue that the greater the content of information in the taxa of a classification and the more characters on which it is
based, the better a given classification will be. More importantly, phenetics
rejects grouping on the basis of *a priori* weighting: no character or type of
character is inherently more important than the rest within all the available
characters of the objects being classified. There is no *a priori* selection of
characters on the basis of some underlying theory. Thus, every character is
of equal weight in locating natural taxa. On the basis of this, pheneticists go
on to specify overall similarity between any two entities as a function of
their individiual similarities in each of the many characters in which they are
being compared. Phenetic classifications are consequently based on all available
characters without any differential weighting of some characters over others.
On these estimates of resemblance, distinct taxa can be recognised because
correlations of characters differ in the groups of organisms under study.

The most noticeable assumption in phenetic methodology is the rejection
of phylogenetic information in classification construction. This is not to say
that pheneticists ruled out all use of phylogenetic information; only that
phylogenetic inferences, while being possible given the taxonomic structure
of a group and certain assumptions about evolutionary pathways and mechanisms, should not be used in reaching taxonomic conclusions. Thus the evolutionary events that produced the observed similarity are not considered,
e.g. questions of evolutionary rates or origins of resemblance.

It is important to note this emphasis on the rejection of evolutionary
assumptions in classification construction; but it must not be taken to mean
that a phenetic classification is necessarily independent of the processes of
evolution. Rather, pheneticists see a distinction between the process of evo-

lution and the process of classification construction. This attitude stems from several related worries over the use of phylogenetic information. Firstly, there is always a lack of congruence between phylogenetic information and similarity estimates. Since the former is always derived from the latter, there is little reason to accept the priority of phylogenetic estimates. Secondly there is a practical problem in phylogenetic reconstruction – phylogenies are largely unknown or at best, imperfect. Thus phenetic classifications are possible for all groups, while evolutionary classifications must be based largely on inferences from present-day organisms. Finally, there are worries over implicit circularity in methods of phylogenetic reconstruction. For example, when no fossil evidence is present, branching sequences have to be inferred from phenetic relationships among existing organisms, implying that there is a circularity when phyletic inferences are based on measures of overall similarity.

These arguments highlight the antimetaphysical element present within phenetics.[7] Gilmour, Sokal and Sneath give a philosophical justification for this attitude, based on views hijacked from Operationalism (e.g. Bridgman[8]) and Logical Empiricism. As Sokal & Camin saw it:

> Operationalism in taxonomy (as in other sciences) demands that state-ments and hypotheses about nature are subject to meaningful questions i.e. those that can be tested by observation and experiment. If we cannot establish objective criteria for defining the categories and operations with which we are concerned, it is impossible to engage in a meaningful dialogue about them.[9]

Not surprisingly, the operational imperative deemed concepts of phylo-genetic homology, blood relationships and species, as well as the principles of evolutionary theory, to be meaningless.[10] In the case of taxonomy, Sneath argued that

> It is often said or implied that taxonomy should be based on phylogeny. The truth is that to all intents and purposes phylogeny is based on tax-onomy . . . [F]rom an estimate of similarity – a taxonomic estimate – I deduce a phyletic relationship.[11]

The point behind such arguments concerns the distinction between what is fact and what is inference. Only secure facts may be employed in classifi-cation construction, so that phenetic information is the only suitable infor-mation. Hence the pheneticists criticise the evolutionary systematists for being non-operational because the degree of inference employed is too great. The phenetic method is seen to be superior because it is empirical[12] – classifications are summaries of observed facts – and objective, because measures of similarity can be re-established by a second taxonomist using a of OTUs. Sneath & Sokal introduced the term 'OTU' to avoid the exclusion of pertinent material from any taxonomic study, while at the same time different set of observations. With the advent of numerical cladistic methods, these attacks were dropped.

So far I have outlined only the principles involved in phenetics. I now want to turn to the actual process of classification. There are four basic steps.[13]

(1) Selection of organisms, groups of organisms or operational taxonomic units (OTUs)

Sokal defines OTUs as

> individuals as such, individuals representing species or higher ranking taxa such as genera or families of plants or animals, or statistical abstractions of the higher ranking taxonomic groups.[14]

In practical terms, the basic units of classification are morphologically homogeneous populations,[15] each drawn from either a single habitat or a single kind of habitat.[16]

(2) Description of the characters that are the bases for comparisons including the estimates of similarity among the OTUs

Characters may be defined as any property that can vary between taxonomic units and the possible values that they can be given are called *states*. It is essential that we compare the same character in different forms.[17] Characters cannot be defined in terms of the 'intelligent ignoramus'[18] as some early pheneticists believed. In addition, character description is non-evolutionary – only estimates of similarity are made. The observed similarity, i.e. OTUs possessing the same state for a particular character, is examined without consideration of the evolutionary event that caused the similarity. The use of assumptions from evolution to justify any inferences about the group are rejected, so that when entities with the highest proportion of shared characters are put together, it is denied that these entities share any particular evolutionary attribute.

The estimation of similarity among OTUs is based on 'overall similarity', and resemblances are calculated using a chosen coefficient (e.g. a simple matching coefficient[19]) and then represented in a matrix of dissimilarity/similarity between OTUs.

(3) Summary and examination of taxonomic structure

Patterns of observed similarity among organisms are used to group and rank organisms in a hierarchical classification. Specifically, this entails the analysis of the constructed symmetrical matrix of such similarity or dissimilarity coefficients using clusters as representations of relationship. When it comes to the definition of groups, pheneticists distinguish between monothetic and polythetic groups.[20] Monothetic groups are formed by rigid and successive logical division so that the possession of a unique set of features is both necessary and sufficient for the group thus defined. In contrast, polythetic groups contain organisms that have the greatest number of shared characters, and no single state in any character state is essential for group membership or is sufficient to make an organism a member of the group.

From Table 1, it can be seen that the class of (a, b, c, d) is fully polythetic because no single character state is found in all four individuals. Consequently, because correlations of characters differ in the groups under study, distinct taxa can be recognised with cluster analysis.

Cluster analysis is a generic term for methods which seek to determine homogeneous subsets of OTUs and produce a classification directly. Two types of cluster methods can be picked out: firstly, simple cluster methods which aim to cover or partition a set of OTUs, and can represent partially overlapping sets. Secondly, there are stratified cluster methods which 'seek systems of clusters in which each cluster has a level, the cluster at each level except the highest being ranked within clusters at a higher level'.[22] Such hierarchic stratified cluster methods produce discrete or continuous clusters at each level.[23] Once constructed, clusters are directly translatable into a phenogram. Phenograms are rooted, branching diagrams which link organisms by estimates of overall similarity, as evidenced by a sample of characters (Fig. 24). The essential property of phenograms is that they are *ultrametric*. Thus, in Fig. 25, the degree of divergence of any pair of populations *a* and *b* is less than or equal to the larger of the degrees of divergence of *a* or *b* from any other population.

Table 1. *Polythetic groups* $+/-$ *indicates the presence or absence or a character* (1–6) *in any given individual* (a–f)[21]

Characters	Individuals					
	a	b	c	d	e	f
1	l	–	+	+	–	–
2	+	+	+	–	–	–
3	+	+	–	+	–	–
4	–	+	+	+	–	–
5	–	+	–	–	+	+
6	–	–	+	–	+	+

So

the
largest
$$d(a,b) < \{d(a,c) , d(b,c)\}$$

If evolutionary rates had been constant, then 'the pairwise dissimilarities of present day populations would be monotone with the time since divergence, and would therefore be ultrametric.'[24] In numerical phenetic studies, then, there is a transformation from a set of pairwise dissimilarities, that are at best metric, to an ultrametric set that defines one or more dendrograms (branching diagrams containing entities linked by some criterion). In proceeding from the ultrametric coefficient to the taxonomic hierarchy, the following inequality holds:

the
largest
$$d(h) < \{d(m) . d(m')\}$$

where h is distance defined, and m is rank defined. Since dendrograms are distance defined, with the vertical scale being ordinal, there is a correspondence with the rank of a classification which functions as an ordinal scale. Thus, both dendrograms and classifications are ultrametric. Dendrograms are distance defined while classifications are rank defined.

(4)   The comparison and choice of classification

The first point to be made is that there are very few phenetic classifications in the form of a completely resolved hierarchical arrangement for taxa, because the phenogram is usually considered to be a classification (Fig. 26). Nevertheless various criteria have been suggested. For instance, Farris[26] describes a general purpose criterion of a phenetic classification as one in which its constituent groups should 'describe the distribution among organ-

Fig. 24 The phenogram.

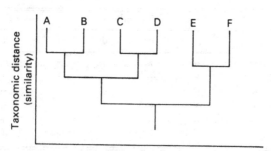

isms of as many features as possible.'[27] Other criteria involve the distortion introduced into the data by the construction of a hierarchic classification, and its representation of phenetic relations. Obviously, the less the distortion, the better the classification.

By way of a summary, the phenetic process of classification can be schematised as follows:

$$
\begin{bmatrix} \text{character state} \\ \text{matrix} \\ (n \times c) \end{bmatrix} \Longrightarrow \begin{bmatrix} \text{similarity} \\ \text{matrix} \\ (n \times n) \end{bmatrix} \Longrightarrow \begin{bmatrix} \text{phenetic} \\ \text{clustering} \\ \text{methods} \end{bmatrix}
$$

c represents the number of characters, with values between 0 and 1.

At all stages in the procedure, the aim is to eliminate bias, especially in the form of evolutionary assumptions. At the same time there is a need for a secure foundation, based on empirical facts and quantitative methods. To many biologists, however, the phenetic view seemed to make the biology fit the maths, and not the maths fit the biology.

The descriptive attitude

The issues which the pheneticists pushed up against concerned the elimination of bias from phenetic procedures in classification construction, and the rejection of explanation, implying that a classification is nothing more than a description supposedly based on empirical findings. It should be clear that there are limits to these attitudes; firstly, bias cannot be totally eliminated from any method. All description has to make appeals to theoretical assumptions; the desire for an unprejudiced classification is a fallacy. Secondly, description can never be fully divorced from explanation.

In phenetic writings both strong and weak descriptive attitudes can be documented. Strong phenetics rejects all appeal to underlying explanation – all notion of hidden cause is rejected – and *ipso facto* argues for the elimination of bias in classification construction. Alternatively, weak (common-sensē)

Fig. 25 The ultrametric property.

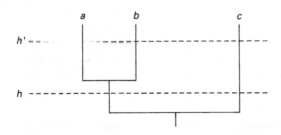

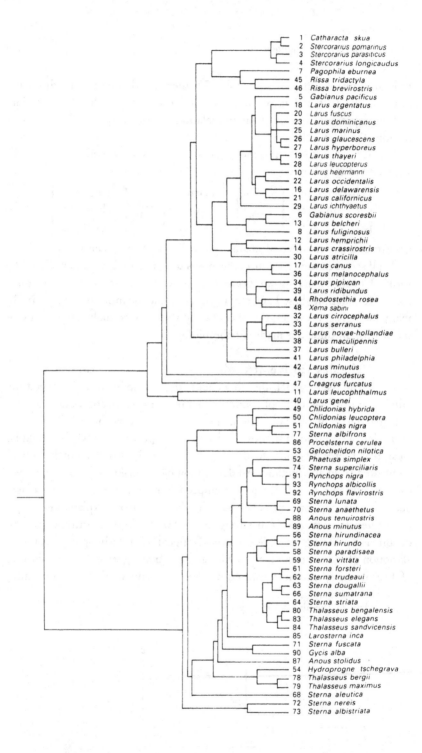

Fig. 26 One phenogram for 81 species of birds. Based on variation of 51 skeletal characters.[25]

phenetics presents a more 'watered-down' version which accepts that bias cannot be totally eliminated, and that recourse to some form of explanation is necessary. Obviously, strong phenetics is a methodologically mad position, implying that resulting classifications are based on monothetic criteria. As such, these classifications are non-empirical, implying that strong pheneticists cannot be practising what they preach. For this reason, the adoption of the aims and methods of weak phenetics is more realistic, since it places greater emphasis on practical problems.

The reasons for the adoption of these positions in phenetics can be traced back to their respective justifications; strong phenetics employs a philosophical justification for the rejection of explanation and wholesale elimination of bias, while weak phenetics voices concern over circularities in methods of classification construction, with the result that its aims and methods are more in agreement.

I will concentrate on the arguments of the strong pheneticists from a largely empirical perspective. In particular, I will examine the requirement of explanatory appeal to a theory, and the rejection of theoretical assumptions in classification construction. I will then go on to explain why the strong pheneticists adopted the attitudes that they did. Since it was not permissible for strong pheneticists to justify their methods through recourse to a theory, the only remaining strategy was to employ philosophical arguments (the external issues that I described in Chapter 1). Finally, throughout this discussion, issues relevant to the theoretical and descriptive attitudes will be contrasted for greater clarification.

The aims of strong phenetics

The attitudes implicit in strong phenetics can be seen in the early writings of Cain,[28] Gilmour,[29] Sokal & Sneath,[30] where motivation for dismissing evolutionary theory is justified philosophically, with the elimination of theoretical presuppositions as an end in itself. Sneath & Sokal also argue for the wholesale elimination of explanatory notions in a classification. In their criticism of evolutionary systematics, they argue that[31]

> By not allowing for the separation between classification and what Michener (1963)[32] has called the 'explanatory element', conventional systematists perpetuate a system which is inherently unstable and hypothetical by the very nature of its operation.

Thus, strong pheneticists argue that the goal of classification is the representation of simple summaries of knowledge, which facilitate convenient storage and retrieval of information. Classifications act as *aides-mémoire*.

Crucial to the strong pheneticist enterprise is the distinction between 'natural' and 'artificial', as first set out by John Stuart Mill and subsequently adopted by Gilmour.

> The ends of a scientific classification are best answered when the objects are formed into groups respecting which a greater number of general propositions can be made, and those propositions more import-ant, than could be made respecting any other groups into which the same things could be distributed. The properties . . . according to which objects are classified, should if possible, be those which are the causes of many other properties; or at any rate, which are sure marks of them . . . A classification thus formed is purely scientific or philo-sophical, and is called a Natural, in contradistinction to a Technical or Artificial classification or arrangement.[33]

The 'objects' or 'things' in biology are species, and the sense of Mill's state-ment is that new information (different types of characters) will conform better to the adopted classification than to another, on the basis of which a wider range of inductive generalisations[34] can be made. For Gilmour,

> A natural classification should . . . be regarded, first and foremost, as that arrangement of living things which enables the greatest number of inductive statements to be made regarding its constituent groups, and which is therefore the most generally useful classification for the investi-gation of living things.[35]

Gilmour's discussion of natural classifications concentrates on two points; first, the search for objects with the greatest number of attributes[36] in com-mon, and second, the overall range of purposes of a classification. In the former, classifications are based on the grouping of objects into classes because of certain attributes they have in common. Those classes containing objects with a large number of attributes in common are known as general purpose classifications, and are useful for a wide range of purposes.

Special purpose classifications consist of classes containing objects with only a few attributes in common, and hence serve a more limited range of purposes. With respect to the latter point, it is stressed that all classifications are purpose/goal orientated because classification is a positive activity. The choice of classification is dictated by one's particular interest in the objects being classified.

> To ask, *in vacuo*, whether one classification is 'better' than another without considering 'better for what purpose?' is an unreasonable ques-tion. Different classifications are required for different purposes.[37]

No classification is ever an end in itself. The aim of a classification must be known so that there is a criterion against which to judge whether one classi-fication is better than another.

It is therefore impossible to have an ideal and absolute classification for any particular set of objects. It is misleading to talk of the *one* classification. One classification may be more natural than another in that there is one factor influencing a group of objects more powerfully than another, so that 'a number of attributes hang together as it were.'[38] Such a classification is useful for a greater number of generalisations.

Gilmour went on to apply these conclusions to biological classifications, by arguing that the difference between biological classifications and those for inanimate objects is one of degree only. 'Living things are an example of objects of which there is one type of classification that is more natural than any other.'[39] Hence, a natural classification in biology is a particular example of natural classifications in general, and *not* a phenomenon peculiar to living things. Gilmour is clearly arguing against the view that the aim of taxonomy is the construction of an ideal classification which reflects phylogenetic relationships. To reinforce this point, he discusses an example based on inanimate objects:

> To say that the obvious, natural classification of furniture into tables, chairs . . . is *made possible* by its manufacturing history is not the same thing as saying that it is actually *based on*, or *expresses* that history, or that it is a historical classification.[40]

While a classification may be made possible by a powerful factor influencing attributes, such a classification may be quite different from one based on that factor, e.g. phylogeny (which would be a special classification). Not only is it impossible to extract one single factor that could be called phylogenetic,[41] but also a natural classification can be constructed without knowing how the factor has worked in influencing the attributes. Speculation based on evolutionary history should not be used. Natural classifications are therefore not the same as phylogenetic classifications, for to suggest that the elucidation of phylogenetic relationships is the goal of taxonomy, implies that taxonomy is an end in itself, when it too, should be purpose orientated. Indeed, the goal of taxonomy is to construct a broad map of the diversity of living things by including as a wide a range of attributes as possible. Taxonomy should strive for a general purpose classification of living things supplemented by any number of special purpose classifications for use in particular fields of investigation.

To summarise, strong pheneticists take natural groups to be based on a larger number of attributes that can be used for a wider range of inductive generalisations. In attempting to distance themselves from the theoretical attitude, strong pheneticists purge the term 'natural' of any monophyletic (i.e. genealogical) import. Since Gilmour naturalness is based on character predictions that can be made from a classification, powerful prediction which admits of objective measurement and testing is all important. In keeping with the descriptive attitude, such classifications must be stable in a manner that is equated with the *repeatability* of results, and hence is the most *objective*. The optimal result, i.e. that which contains the maximum information content about the classified groups, is as good as any other. As Sneath & Sokal see it,

> . . . [with] the greater the content of information in the taxa of a classification and the more characters on which it is based, the better a given classifiction will be.[42]

Since there is supposedly no theoretical underpinning in strong phenetics, stability at the level of the binomial nomenclature and higher taxa is crucially important in judging a classification. For strong pheneticists the stability of a classification can be affected in one of two ways; first, by an increase in information content (i.e. new characters) so that as the number of characters is increased, the classification tends towards a limiting ideal classification. Jardine & Sibson have succinctly argued in this respect, that the pursuit of an optimal classification is impossible, and that the stability of a classification 'is entirely a matter for empirical investigation.'[43] Such a conclusion is not surprising when it is realised that as the number of characters increases, then there is no decrease in change in relative similarities. Second, through the inclusion of a new taxonomic entity in subsequent studies. Because stability cannot be measured against a gradual improvement based on approaching the true phylogenetic relationship, it must be measured through predictive power and information content. Here, prediction is the degree to which a specific classification agrees with characters not used in the formulation of that classification, and as such, the classification must be based on overall similarity because overall similarity gives greater predictive power.[44] In the case of information content, we have already seen that a Linnaean classification is nothing more than a system of names hierarchically arranged, so that classifications are first and foremost devices to store and retrieve information about organisms. The information content alluded to resides in the logical structure of hierarchical systems in general. Information is the result of objects being classified into sets, while the information content of a classification is only these hierarchical sets; nothing more can be retrieved. Thus only groups are counted; the content of such groups is ignored (see Chapter 5). All cladists have been quick to pick up on this point, arguing that the stability of a classification is under differential selection of characters, so that the content of the groups must be included in a measure of information content.[45] Clearly, the notion of information content adopted by all pheneticists is somewhat redundant, since the only real way that stability can be measured is through predictive power. Predictive power and stability are one and the same thing.

It cannot be denied that the aims of strong phenetics are commendable, with their emphasis on utilitarian views based on convenience and prediction. Such views exemplify the descriptive attitude. But are these aims sensible? Should classifications be nothing more than tools for serving predictive purposes? Do the methods of strong phenetics give maximum predictive power? Before confronting such questions, I want to situate phenetics more closely into the context of the descriptive attitude, so that the limits of the descriptive attitude, as discussed in Chapter 1, can be applied to strong phenetics. Specifically, our concern will be with the 'uncoupling' of description and explanation, as encountered in the descriptive and theoretical attitudes. This discussion will lay the ground work for the subsequent analysis of the aims and methods of strong phenetics.

The uncoupling of description and explanation

In the taxonomic literature the status of scientific vocabularies (including classification) is discussed in terms of description and explanation.[46] Farris views scientific vocabularies

> not only as a means of description, but also of discussion and explanation of the natural processes that give rise to observation.[47]

For the descriptive and theoretical attitudes this represents an area of tension because the former demands the uncoupling of description and explanation.

In the descriptive attitude, as represented by phenetics, distances between objects are measured and then converted into a distance matrix based on measured similarities and/or dissimilarities. This matrix is then fitted to an ultrametric matrix using a method based on 'structural fit', and subsequently converted into a classification. In this method no evolutionary assumptions are made, so that we are searching for a best fit of the distance matrix to the ultrametric matrix (Fig. 27). But, there is a problem here; on what basis do we choose the best fit? In the absence of any theoretical expectation this is an impossible task because no justification for the choice of fit can be given.[48] A simple example will clarify what I mean. Given a set of data, such as a matrix of four pairs of distances, it is assumed that this data ideally conforms to a certain structure. For instance in the co-ordinates

0.95,0	1.1,1.05
0,0	0,1.2

it is hoped that they conform to the model of a square, given by the following co-ordinates:

1,0	1,1
0,0	0,1

Clearly the data closely conform to the model of a square, and any deviation from the model would technically constitute error.

Obviously, the measurement of error and choice of best-fit becomes more complex if we are searching for a tree structure which has the best-fit to the data. No justification for the choice of best-fit can be found, unless some assumptions about evolution are included. The essential point here is that there must always be some recourse to explanation if we are to justify our methods. There must be some underlying explanatory justification for our choice of method.

While the descriptive attitude incorporates a structural model in which distances between objects are fitted to some model, in the theoretical attitude there is model of process based on evolutionary change. The aim is to represent evolutionary events by fitting the data into a tree pattern which should conform to a particular model of change, such as in the Additive model which depicts divergent change. Phylogenetic cladists such as Eldredge & Cracraft[49] follow this method by first deducing tree structures and then fitting them to a model of process. Naturally enough, some changes do not conform to the model of additive change, implying that errors are present. These errors are associated with the branch lengths of the tree because the model assumes that the evolutionary process is divergent. Any cases of undetected convergence will be represented as errors in the branch lengths (Fig. 27). As in the descriptive attitude, there is a problem over how to choose the criterion of fit of any model to the tree pattern. When should we reject the fit of the additive model onto the tree pattern? Perceived error can either be ignored, or explained away in terms of the evolutionary process. In the latter, the need for an estimation of error implies that much more needs to be known about evolutionary history than is actually the case at present. However, phylogenetic cladists who believe that evolution is divergent will not attempt to discover instances of convergence, and any error encountered in the 'fit' of the tree pattern to their model will be ignored. A fixed picture of evolution will be presented, and all counterinstances will be explained away as faults in the method, when really it is the model that is at fault. Failing this, the model should be used to test the methods. Alternatively, if the model of evolutionary change is rejected, then recourse to the descriptive attitude may not necessarily help. Thus, both the extremes are confounded over a failure to justify the best-fit of the data to either a model of evolutionary change or to a structural model. While it may be desirable to uncouple explanation and description, in practice it is not possible. Both the theoretical and descriptive attitudes are confounded over attempts at teasing these two aspects apart.

The elimination of bias

Given that bias cannot be eliminated from methods of classification construction, I want to examine the ways in which strong phenetics fails to remove bias. I will concentrate on two aspects: character description and taxonomic structure.

(i)   Character description

In character description, bias can enter in at several levels. Firstly, there is the problem of recognising the basic characters in any study. In scoring the characters of organisms for comparison, we must decide what is the 'same' character and the 'same' state in two forms. In many instances this will be no problem, but in others, especially if taxa are distantly related, there will be uncertainty. Sokal & Sneath advocate the use of operational homology to decide whether or not two characters are the 'same'. Thus, two character states are deemed the 'same' whenever they are indistinguishable. But, as I argued in Chapter 1, there is practical ineliminability over the recognition of similar characters in homology. The use of operational homology presupposes a model for decision making based on some criterion of recognition.

Similar problems over the introduction of bias arise in the characterisation

Fig. 27 The theoretical and descriptive attitudes contrasted. (*a*) The descriptive attitude is based on a structural approach. (*b*) The theoretical attitude is based on fitting data to an additive structure. But problems are encountered in distinguishing between branch lengths and their respective errors. *x* = branch length, *r* = error.

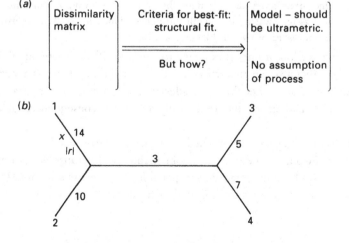

employ[ing] taxonomic units of different categorical ranks as the
entities to be grouped into more inclusive aggregates during classification.[50]

OTUs can therefore be 'individuals, exemplars of genera, or averages representing species'.[51] Problems arise frequently in supraindividual categories
where a taxon used as an OTU proves to be variable for one or more characters. In higher taxa, the majority of characters vary. Consequently, some
choice has to be made over how to treat such characters, and this will necessarily introduce bias. For example, Sneath & Sokal[52] mention the *exemplar*
method in which a single representative of a polymorphic group is an exemplar of the taxa it represents. Not only does this method assume that the
variance of the polymorphic forms within their taxon is less than the variance among the taxa of the study, but also the use of a single representative
of an OTU as an exemplar of the taxa it represents, introduces the notion of
a hypothetical average OTU. This has theoretical connotations. Thus, while
the selection of OTUs is supposedly based on empirical criteria, in practice,
such criteria are non-operational.

Further problems in character description and taxa formation occur in the
meaning of overall resemblance. Cain & Harrison[53] originally defined affinity
as the degree of overall resemblance estimated from all available characters.
Phenetics was seen as

> the arrangement of overall similarity based on all available characters
> without any weighting.[54]

Phenetics was contrasted with phyletics, 'the arrangement that aims to show
the course of evolution'.[55] Central to the phenetic approach was an *analysis of
judgement*, which involved the

> summing up in one way all the differences between any two forms at a
> time, irrespective of what units they are expressed in, and in using the
> total as his measure of overall difference, which is readily convertible
> to a measure of overall likeness or affinity.[56]

This strong version of phenetics aimed to base overall similarity on all characters, such that a single character was taken as anything that was a variable
independent of any other thing considered at the same time, i.e. characters
that are not mathematically or logically necessary consequences of each
other.[57]

But overall similarity as implied in Cain & Harrison's formulation is
impossible. Strong phenetics is an unobtainable ideal because we can never
be sure that we have the totality of all possible characters in a set of OTUs.[58]
In practice we are always dealing with a finite set of characters. Nor is it

possible to have an absolute measure of similarity (i.e. matching correspondence) between non-identical sets of characters which are infinite. It is impossible to specify any parameters,[59] so that our measure of overall similarity would be unbounded and unconstrained.

The formulation of overall resemblance given in strong phenetics must be weakened. Phenetics should be founded on relationships of similarity which are assessed on the basis of a restricted universe of characters. At this point theoretical bias arises because from an evolutionary perspective, we are dealing with characters that have come to differ between OTUs since the latter diverged from their most recent common ancestor. Since ancestors are never known with certainty, the crucial difficulty lies in recognising when one has, in fact, actually obtained this finite set of characters. Even without this evolutionary attitude, it would still be difficult to justify that a finite set had been obtained without invoking some theoretical bias. So, while on the one hand, the goal of a general classification based on all characters seems untenable, on the other hand, any weakening to a finite number of characters introduces bias, because however many characters one lists, a subjective selection is made from the totality of characters of the individuals, thereby injecting a subjective element into the process right at the start, and complicated by possible character incongruity. In addition, there is an implicit restriction in the basis of relationship to include only unweighted or equally weighted characters.

The third area of tension which arises in strong phenetics concerns the domain from which the characters are to be chosen. Strong pheneticists assume that phenetic similarity ought to be obtained from the phenotype. According to Dobzhansky

> the phenotype of an individual is what is perceived by observation:-the organism's structure and functions – in short, what a living being appears to be to our sense organs, unaided or assisted by various devices.[60]

Strong pheneticists argue that all observable characters should be employed, yet in some instances this is taken to include genetic evidence.[61] Leaving aside the difficulties in qualifying observable characters, it is still misleading for strong pheneticists to distinguish between phenotype and genotype when arguing for an objective classification based on empirical criteria. Furthermore, strong pheneticists tend to concentrate on particular kinds of data viz. morphological, despite the fact that other kinds of data (e.g. behavioural) are equally valid.[62] Such resultant classifications can be nothing more than artificial in the sense of Gilmour.

Conclusively, in the realm of character selection and description, strong pheneticists cannot uphold the standards that they have set themselves. The desire to purge all subjectivity from character analysis is impossible.

(ii)   *Taxonomic structure*

Given that strong pheneticists aim for predictive power and objectivity in their classifications, does the Linnaean hierarchy adequately satisfy these requirements? Is it the best data structure for their purposes? Despite the general observation that a hierarchical classification into groups is possible because of theoretical confirmation from evolutionary theory, the strong pheneticist believes that hierarchical classifications are best suited for economising memory, i.e. convenient storage and retrieval of information. However, there is not necessarily a good case for the strong pheneticist to advocate hierarchical classifications. A comparison with non-hierarchical systems will bear this out.

In a series of papers Dupraw[63] has set out some of the aims and methods of non-hierarchical classifications.[64] His methods do not produce classifications in the sense implied normally, but give ordinations obtained by principal component analysis.[65] For Dupraw, the

> Primary goal of non-Linnaean taxonomy is the recognition of individual organisms (as opposed to defined categories of organisms); its justification is the unrestricted transfer of biological information from the known specimens which yield the information, to the populations of unknown specimens which they represent.[66]

Recognition of an unknown specimen in the system is accomplished by interpreting its coordinate position[67] in the classification relative to various positions of known specimens in the same classifications.

> In essence, the classification and its coefficients are the results of a multiple-group discriminatory analysis based on sample measurements from known specimens. The effect of the discriminant analysis is to provide an optimum view of the patterns of variation in the known specimens, and this pattern has predictive value for any unknown specimens subsequently compared with the known.[68]

Dupraw argues on several counts that non-Linnaean classifications lead to the removal of theoretical presuppositions implicit within Linnaean classifications.

Firstly, there is the introduction of serious information distortion through the use of categories in the Linnaean hierarchy. Informational artefact occurs in cases where the biological relationships are not mutually exclusive, and though they may have an 'evolutionary nested structure', generally they do not occupy hierarchical levels. Sokal & Sneath were aware of this and commented that

In devising our methods in such a way as to avoid overlapping clusters, we are in fact biasing the data to yield discrete definable characters.[69]

Distortion occurs because of the high similarity of specimens just to either side of the category boundaries and the dissimilarity of extreme specimens compared within the same category boundary. Contrastingly, non-Linnaean classifications are direct functions of the properties of individual specimens and provide direct 'views' of the patterns of individual variation, as measured in a set of continuous variables. Categorisation is entirely omitted (cf. Sattler's views in Chapter 1).

A second source of bias for Dupraw arises in the structuring of a classification to represent theoretical relationships. The incorporation of theoretical relationships can only diminish the objectivity and usefulness of the classification as a medium of communication. To the extent that a taxonomy depends for its validity on outside areas of theoretical biology, its classifications become subject to alterations with every theoretical advance, thereby reducing stability. Dupraw concludes that when the goal or function of classification is defined in terms of largely hypothetical relationships among specimens, it becomes impossible to find an objective basis for judging the success of alternative classifications.

Finally, the aim is not to transmit any information on phylogeny but to provide for the transfer of all kinds of information by associating samples of known specimens to the population of unknown specimens. Good information transfer implies good prediction. In hierarchical systems, however, the classification is usually built around a specific kind of information, e.g. phylogenetic, genetic, etc. Such theoretical structuring decreases the recognition function. Only the transfer of general information is best achieved by efficient methods of specimen recognition, since no prior and unspecified judgements are made.

Strong pheneticists would agree with Dupraw over this final point, and view it as a criticism of theoretical attitudes in general;

> The fundamental weakness of phylogenetic systematics is well known; hypotheses about phylogenetic relationships are used as evidence about taxonomic relationships, which in turn yield judgements concerning the phylogenetic relationships of other structures and forms.[70]

Likewise, over the second point, they would argue that their classifications are general purpose classifications which do not involve theoretical relationships. But, with respect to the distortion of information through categorisation, strong phenetics is certainly open to this charge. They cannot claim that their system is fully empirical and without bias in the face of non-Linnaean systems providing a more empirical approach on the basis of removing distortion in categorisation, and in not presupposing any particular type of information. Indeed, if strong phenetics is to be internally consistent it must justify the use of hierarchical systems on grounds other than facilitating the representation of phylogenetic hypotheses. It must be shown that hierarchical systems do not necessarily presuppose covert evolutionary assumptions, and are necessary for other reasons. As we shall see, the strong pheneticist's rejection of points two and three above is not justified, despite considerations over the storage and retrieval of information and predictive power in classifications.

Some of these considerations are discussed by Jardine[71] in his comparison of hierarchic and non-hierarchic cluster methods, where a 'dissimilarity coefficient on a set of populations is transformed into a classification.'[72] In hierarchic (stratified) cluster methods the aim is to seek systems of clusters in which each cluster has a level, the clusters at each level except the highest being nested within clusters at a higher level. All clusters are discrete. In non-hierarchic (simple) cluster methods the aim is to find a simple covering or partition of a set of populations (i.e. clumping). Major differences in hierarchical and non-hierarchical systems arise in the storage and retrieval of information. Hierarchical systems have a privileged position as a mode of classification because of good information retrieval. The hierarchy owes its usefulness as a data structure for the retrieval of information. In contrast, non-hierarchical systems are bad data retrieval systems. Dupraw argues against this view, believing instead, that non-hierarchical systems provide more accurate information retrieval. In his non-hierarchical system, increasing precision in identification is possible due to multivariate statistics, in which all the variables of the analysis can be considered simultaneously. False weighting is therefore avoided. Furthermore, given that the power to associate specimens which are the same in many or all of their attributes is due to the recognition of specimens, then undescribed forms are more likely to be recognised and inferences can be made regarding their affinities. In a hierarchical system (e.g. the Linnaean system) specimens are employed to define sets of mutually exclusive categories, and unknown specimens are recognised by deciding the categories to which they supposedly belong. Consequently, a dichotomous key is incapable of dealing with 'undescribed' species which were not available when the key was constructed. Such undescribed forms are usually sorted (falsely) into one or other of the *known* categories which the key was built to distinguish, and there is usually no reason to expect even then that they have been associated with the known specimens they most nearly resemble.

Dupraw's arguments are misleading, because while non-hierarchic methods will increase the accuracy of representation, they will at the same time increase the complexity of the resultant clustering. Retrieval of information cannot be executed conveniently. There is another reason for doubting Dupraw's views on retrieval of information; the term 'recognition' as used by Dupraw is not the same as the term 'identification'. Dupraw is really emphasising the good predictive power of non-hierarchical systems, which is a separate matter from the storage and retrieval of information. It is here that non-hierarchical systems come into their own, by enhancing predictive power.

Thus non-hierarchical systems have high predictive power with respect to the addition of unknown specimens. Indeed, the confidence and information associated with a non-hierarchical system increases in proportion to the number of specimen points it contains. This follows from the fact that such classifications are derived from a statistical/probabilistic procedure (in the sense that biological classifications are based on limited temporal and spatial examples of the objects to be classified). Non-hierarchical classifications therefore satisfy Gilmour naturalness better than hierarchical systems, because Gilmour naturalness advocates predictive power and clear association between unknown specimens and previously discovered specimen groupings. Furthermore, the more generalisations that can be made from the groupings the more natural the classification. In this respect, the most natural classification would be the one that most closely reflects the specimen distribution in a high dimension character hyperspace (since the nearer two specimen points lie, the more generalisations can be made about them in common and vice versa).

For Dupraw, since recognition is the associating together of objects which are the same in many or all their attributes, it follows that the most natural classification (sensu Gilmour) is the one which best provides for the recognition of specimens. The principle of specimen recognition is therefore logically inseparable from the notion of naturalness. The recognition power of a classification is the most objective criterion for its naturalness.

To summarise, non-hierarchical

> classifications set out to predict nothing except the pattern of variation
> in the measurable properties of the postulates.[73]

In this way, they exemplify Gilmour naturalness more accurately than hierarchical systems. Non-hierarchical systems are also more stable. This follows from the point made earlier that in the absence of an underlying causal theory in the descriptive attitude, stability and predictive power are one and the same. In hierarchical systems with the creation of new taxa and their assignment to various categories, there is no *a priori* reason why the stability of a classification can be guaranteed under the extension of the domain of objects, or through an increase in the number of characters used. In the non-hierarchical system there is. So while hierarchical systems are better storage and retrieval systems, they lack predictive power. There are also difficulties in non-hierarchical systems; a compromise has to be made between complexity and accuracy of representation. Greater accuracy implies greater complexity, and hence greater difficulty in convenient retrieval. Despite these difficulties, if strong pheneticists are advocating Gilmour naturalness, then it is dishonest not to reject non-hierarchical systems only because of their lack of convenience and economy of memory. Indeed, if strong pheneticists want to show that their classifications are empirical and objective, then the direct projection of individual specimens in character hyperspace gives increased objectivity and repeatability to the classification. Over and above this general criticism of not practising what they preach, strong pheneticists are also guilty of failing to remove or point out theoretical assumptions that are 'wedded' to the Linnaean hierarchy. In any hierarchical system there is always the underlying assumption of homogeneity of taxa, that all groups included at a particular hierarchical level are similar in kind of nature. Such homogeneity represents a loss of information and introduces distortion because the similarities between populations based on many attributes usually shows imperfect hierarchical structure. The assumption of homogeneity implies that there must be some underlying process which does not give random patterns, but patterns that are discrete and hierarchical. Yet, if the interest is in describing patterns, it is odd that the strong pheneticists have such an obsession with hierarchical retrieval systems. Strong pheneticists do not even discuss the imposition of hierarchical structures on random data. But such tests are possible since the extent to which similarities/dissimilarities depart from a hierarchical structure may be conveniently measured by the distortion imposed by the subdominant ultrametric. From an evolutionary perspective, the ultrametric inequality assumes that all character changes occur simultaneously for all time and place. Without hierarchical systems, patterns of variation can be more easily represented, without necessarily presupposing any particular evolutionary (or other theoretical) assumptions, or ignoring examples of reticulate evolution due to inbreeding (especially in plants). The use of hierarchical systems cannot be consistent with a descriptive attitude that places special emphasis on prediction.

In the realm of character description and taxonomic structuring, the strong pheneticists fail to live up to the standards that they have set themselves. They do not practice what they preach. Strong pheneticist views, as typified in Sokal & Rohlf's 'intelligent ignoramus',[74] encounter theoretical bias and presuppose theoretical assumptions (not necessarily evolutionary). In conclusion, the strong pheneticists are dishonest in retaining the Linnaean hierarchy, since much of what they require can be better obtained using non-hierarchical systems.

The philosophical justification

Given that the strong pheneticists do not practice what they preach, it now remains only to discuss the manner in which their aims are supposedly justified through the philosophy of science. Within the writings of Gilmour & Walters there is constant appeal to positivist attitudes, while in Sokal & Sneath there is further emphasis on operationalism. The role of philosophy is largely prescriptive; we should imbibe our aims and methods with the beliefs of a particular philosophical perspective. In this case, because strong pheneticists prescribe positivist attitudes, then classifications should be objective and enhance prediction. The pursuit of such an ideal derived from philosophical argument is fair enough, but if inconsistency is to be avoided, then there must be some correspondence between what actually occurs in classification and what should happen. This amounts to the simple description of our methods and aims, and subsequent prescriptions for improving them. But strong pheneticists ignore any interplay between description and prescription; they employ philosophical views for prescriptive purposes, paying little attention to whether their aims and methods are, in the first place, concordant with positivist attitudes.

In his discussion of natural and artificial classifications, Gilmour incorporates an epistemological analysis to examine the tenets of classification.[75] This analysis is derived from Dingle's discussion of the rational and empirical elements in 'the objective world of physics'.[76] The empirical element is the given; the 'unalterable' sense data, which must be 'clipped' together in a rational relationship. For Gilmour, this means that

> The classifier experiences a vast number of sense data which he clips together into classes, each of which is definable in terms of specific data.[77]

Classification includes rational and empirical elements, as expressed in 'the countless sense data of experience out of which reason builds that logically coherent pattern which we call the external world.'[78] Thus the purpose of classification is to enable the classifier to make inductive generalisations concerning the sense data he is classifying.

Two further points should be added. Firstly, there is an analogy between the classification of objects and sense-impressions. Gilmour & Walters[79] argue that since objects are grouped into classes because of certain attributes they have in common, then a comparison can be made below the level of objects where individual objects themselves are the result of classifying our sense-impressions into classes that we call individuals. Thus, the individual is a concept, the rational construction from sense data, and the latter (units) are the real, objective material of classification. Secondly, Gilmour argues against the view that natural groups, such as species rank and lower, have a 'biological reality' in the process of evolution which artificial groups do not possess. Because both natural and artificial groups are concepts based on experienced data, 'reality' here is not 'metaphysical objectivity' (i.e. an objective, outside world which exists independently of the mind). Instead the difference between artificial and natural classifications, is over the number of attributes and overall range of purposes. Both types are created by the classifer for the purpose of making inductive generalisations regarding living things.

In these respects, there are superficial similarities with logical positivism.[80] For example, there is great emphasis on observation; what we can see, etc., provides the best foundation for the rest of our knowledge. There is also a distrust of theoretical entities; reality is restricted to observables. Closely linked to this attitude is a rejection of all causality in nature, over and above the constancy with which events of one kind are followed by events of another kind, and the rejection of all non-scientific explanations. Finally, there is emphasis on verification. Significant propositions are those whose truth or falsehood can be settled in some way. For the strong pheneticist, operationalism plays an important role here; all physical entities, properties and processes are definable in terms of the set of operations by which they are apprehended.

However, on closer examination, the issues being confronted by the strong pheneticists are not really positivist, despite positivist overtones in Sneath & Sokal about sticking to the data:

> A basic attitude of numerical taxonomists is the strict separation of phylogenetic speculation from taxonomic procedures. Taxonomic relationships are evaluated purely on the basis of the resemblances existing *now* in the material at hand.[81]

In the first place, positivism calls for a neutral observation language, with the formulation of theories as axiomatic calculi co-joined with observational terms via bridge hypotheses. There is a strict separation between theory and observation. But for the strong pheneticist the plea for a neutral approach is motivated through a questioning of theoretical assumptions. Just requiring as few assumptions as possible is not positivist, and some of the assumptions that are called into question, e.g. natural selection, would be retained by positivists. The rejection of all theories by strong pheneticists is not the same as scepticism over theoretical entities, and their means of verification. Positivism still requires theories.

Strong pheneticists may disagree with this; they could argue that an extensive and elaborate classification may be constructed in isolation from all scientific theories and transformed, only later, into a theoretically significant classification. In this sense, the relation between evolutionary theory and classification is the same as the relation between theory and observation.[82] Classifications are nothing more than observational devices. This is clearly misguided; not only are classifications polythetic, but the sense of a classification as a neutral observation language is untenable, because classifications are hierarchical and so distort the data. There are other reasons for the inadequacy of this analogy. For example, the observation level in the historical sciences, such as paleontology, is very far removed from the actual event. There is a lack of resolution over individual events, since it is impossible to talk of the one speciation event, or the particular ancestral–descendant relationship which gave rise to a new form. Instead, reference is made to speciation events in general; that is, to low-level generalisations. In contrast, in particle physics, it is quite plausible to talk of electron traces as being observables. Such individual traces are the observed, and are extrapolated into general schemes. But in the historical sciences there is no equivalent to an electron trace; this story (i.e. electron trace) can never be extrapolated to these phenomena. On this basis a classification has facets in common with a low-level theory, in which the units of classification double as observational and theoretical terms. For instance, the term 'population' is both theoretical and observational; it can refer to populations through time, or to the distribution of characters of these organisms. But in a classification we can never refer to this population. Thus classifications and evolutionary theory do not exhibit strict observation–theory relationships as pheneticists might require in advocating positivism.

Secondly, the phenetic use of operationalism, where statements and hypotheses about nature are subject to meaningful questions, i.e. testable through observation and experiment, falls short of the mark. Attempts to yield information simultaneously on the degree of resemblance, descent and rate of evolutionary progress would lead to excessive complications under operationalism, and are therefore rejected by pheneticists. But the danger of operationalism, if carried too far, is it leads to inference-free direct observation which is unlikely to lead to theoretical advances.[83] Nor is it true that observation, in the context of phenetic methods, is in any way direct. Problems of using an operational criterion in homology are encountered due to the practical ineliminability of theoretical assumptions.

Finally, strong pheneticists require objectively the most useful classification. They dream of an unprejudiced classification. But while the positivist interprets 'objective' as implying a neutral foundation upon which all inferences can be verified, the strong pheneticist argues that objectivity is equivalent to repeatability on the computer. To talk of objectively the most useful classification in this sense is certainly not positivist.

Clearly, there is a difference between the descriptive attitude manifest in strong phenetics, and the views of positivism. In the former, there is the rejection of theoretical notions, with the specification of classifications as tools for serving some function, e.g. prediction. In Gilmour, classifications are seen as ways of retrieving and/or throwing away information, and represent innumerable ways of simplifying data. Classifications are descriptions, and are not concerned with knowledge claims as such about the one true world (see Chapter 5) since different classifications can be obtained by different methods.[84] In positivism, however, there is a real concern over making truth claims about the world; hence their concern with epistemological claims about how we can come to know the outside world (which the strong pheneticists also adopt, but for more instrumentalist reasons). While the strong pheneticists may advocate epistemic objectivity, i.e. objective methods for the construction of classifications, they are at the same time detailing a move away from all theoretical considerations about the 'real' world. Logical positivists never took such an extreme view: they only rejected theoretical propositions which could not be verified empirically or logically.

In conclusion, the so called positivism espoused by the strong pheneticists is really a side issue; it is not a true reflection of the descriptive techniques employed, and is of no significance *in practice*.[85] Justification from positivism has only led to the prejudice of neutral data and methods, giving rise to an extreme and incoherent position, based on idealistic aims. Cries of 'data alone' have led to a good deal of muddling over not necessarily importing hypotheses and necessarily not importing hypotheses. Nor is it possible to achieve the ideal of an impersonal, objective method because there must always be some appeal to explanation, and hence the incorporation of theoretical bias. Bias occurs in what the classification represents with respect to assumptions made during its construction, as well as in implicit assumptions in 'pure' description. Given these problems, it is not surprising that phenetics underwent changes which made the aims and methods more concordant. I will briefly outline some of these changes.

Weak phenetics

After the early enthusiasm of the strong form, a more stable, weaker version arose.[86] This 'common-sense' phenetics has a more practical approach to the descriptive attitude, so as to counter problems of inference in the scenario approach (historical narratives which attempt to describe which groups give rise to which, as well as the ecological changes which produce the adaptations that characterise the organisms). Specific doubts over particular aspects of Darwinian evolution are raised, and various rules of thumb, based on uniform direction, rates of evolution, irreversibility of evolution are rejected. This weak version of phenetics retains the basic element – the estimation of relationship due to similarity, but effectively does away with the idea of an objective data base treatment. Only a restricted subset of characters is used, and there is a relaxation in the specification of unweighted phenetic relationships. Characters can either be equally weighted (e.g. through accident or by choice) or differentially weighted. In the latter, weighted phenetic relationships can be found either by treating the data as a result of the individual performance of characters within the study (e.g. by deleting or downgrading characters that are highly variable or strongly correlated with other characters); or by differential data treatment resulting from hypotheses and rules derived from criteria external to the study, and usually related to the beliefs of the worker, e.g. characters upgraded due to their relative importance, complexity etc. Such attitudes are possible because the motivation for this new approach is based on worries over circularity of inference (not in the crude form first proposed by the strong phenticists[87]) and underdetermination of the theory by the data:

> . . the proponents of numerical taxonomy . . . are firmly convinced that phylogeny is responsible for the existence and structure of the natural system. They are criticizing not evolution nor the study of phylogeny but speculation passed off as fact.[88]

Two points follow from this weakening in phenetics; firstly, it becomes clear that strong phenetics is really the weak version dressed up in philosophical terminology. Secondly, with the loosening of rules, there is more of an opportunity for phylogenetic influences to be smuggled into phenetic relationships. Farris's criticism[89] of phenetics marks the end point of this trend. Farris argues that for a classification to be natural, it must be a reflection of phylogeny, and hence evolution. Phylogenetic cladistics can give more natural classifications, in the sense of Mill and Gilmour, than phenetics, because the former expresses greater information content. In Farris, this follows from the fact that not all features of an organism (as in overall similarity) are equally informative, and the methods of phylogenetic cladistics estimate those characters which are more informative (using special similarity – similarity assessed on the basis of synapomorphies alone), and hence yield a more natural classification.

Superficially, this appears a convincing argument, but on closer examination, there are several misunderstandings. Farris interprets Gilmour naturalness as meaning that each group recognised in a classification should be characterised by at least one character state unique to that group, i.e. a monothetic criterion.[90] But this criterion is not a corollary of Gilmour naturalness, where emphasis is on the distribution of all character states, whether 'primitive' or 'derived' so that a classification is characterised as polythetic. In addition, Farris uses a concept of synapomorphy that is not the same as used by most cladists, and it is even debatable whether such a concept manages to represent plesiomorphy (primitive character states) and apomorphy (derived character states). Farris has only provided an easy knockdown argument, which in the light of close analysis, is based on standards that are relevant to his views and not to those of the pheneticists.

5

Transformed cladistics and the methodological turn

Introduction

In 1979 Norman Platnick[1] published a paper entitled 'Philosophy and the Transformation of Cladistics' which set out some of the basic principles of the new taxonomic school and their derivation from classical Hennigian cladistics. There followed a flurry of publications, e.g. Nelson & Platnick[2], Patterson[3], which attempted to put transformed cladistics on a more secure footing. The fact that transformed cladistics gained the centre stage of the present-day controversy in taxonomy represents a testimony to the success of the enterprise. Quite whether such a standing is justified is questionable. Unfortunatley, the few criticisms that have been levelled at transformed cladistics have either been misguided or misinformed (e.g. Beatty[4], Charig[5], Ridley[6]). Further to this, disagreement and misunderstanding have arisen from the transformed cladists' lack of attention to issues that some of the other schools regard as central to any discussion in taxonomy. The need for finding a common ground between transformed cladistics and other schools is therefore important.

Certainly, part of the confusion is a product of the historical development and its similarity of method with phylognetic cladistics. What these similarities are, and where the differences lie must be examined closely since the dividing line between the two is not always explicit. By way of introduction I will first summarise the differences between phylogenetic and transformed cladistics before turning to the procedures that transformed cladists utilise in classification construction.

As we have seen, early cladistic views concentrated on the reconstruction of evolutionary history, while later, more general views of cladistics attempt to discern the natural order in any system that involves some sort of descent with modification.[7] Simpson[8] saw the shift as being based on canonical and non-canonical versions of cladistics, i.e. whether their proponents do or do not regard cladograms as equivalent to phylogenetic trees. However, this distinction does not separate out adequately the differences between trans- formed and phylogenetic cladistics. Panchen[9] suggests that the differences lie in the types of cladistic analysis: one limited to historically developed patterns, the other, a more general notion, applicable to all patterns. In the latter respect, Nelson[10] has produced a method of component analysis, a general calculus for discerning and representing patterns of all sorts. This is more accurate, since it emphasises the possibility of ignoring all time based phenom- ena, and concentrating on patterns in general.

Transformed cladistics represents the logical end point of a move away from representing the particulars of evolutionary development in classifi- cation. For Platnick

> The implication is that cladistic methods are not the methods of
> phylogenetics *per se*, but the methods of taxonomy in general, and that
> our knowledge of phylogeny stems from our knowledge of taxonomy.[11]

Indeed, 'good taxonomy' is not a tradition of a particular theory of evolution but a tradition of looking for unambiguously defined groups of taxa. Patterson[12] further argues for the lack of any necessary connection between phylogeny and systematics.

Transformed cladists aim to discover the 'natural order' by searching for patterns in nature, so that

> A cladistic classification . . . is one in which the taxa that are
> recognised are equivalent to the groups united in branching diagrams,
> or cladograms, by the distribution of unique characters, and the expec-
> tation that these characters are just representatives of a general
> pattern.[13]

The natural classification of a group is the one which is supported by the greatest number of characters, thereby identifying patterns which may or may not have any biological significance attached to them. Transformed cladists also argue that since we have no knowledge of evolutionary history which is independent of the existence and discovery of a natural hierarchic system, the investigation of patterns in the distribution of characters (tax- onomy) is both independent of, and a necessary prerequisite to, any investi- gation of evolutionary processes. This distinction between patterns (aspects of the orderliness of life) and processes (mechanisms that generate these patterns) is crucial. This will be examined in detail in the next chapter. For the phylogenetic cladist process is superimposed over pattern – the pattern is necessarily phylogenetic because it is the product of descent with modification.

'The expected outcome is a nested set of evolutionary resemblances.'[14] In contrast, transformed cladists do not superimpose process over pattern, since their major concern is over the *recognition* of patterns. Contrary to what might be expected, there is no argument between phylogenetic and transformed cladists over how the groups are recognised, nor over what the groups contain. The difference lies in the phylogenetic cladists' need to appeal to the evolutionary process for the justification of these groups. Thus Wiley[15] states that 'characters alone are insufficient to define a natural taxon.'

To understand best the methods of the transformed cladists, similarities and differences with phylogenetic cladistics must be emphasised. In general, transformed cladists increase the emphasis on methodological rules in an attempt to reject all appeals to an evolutionary explanation. Their methodology reduces the evolutionary components in classification construction, e.g. the distinction between pattern and process. Closely associated with this distinction is the one made between cladograms, trees and scenarios.[16] Cladograms are viewed as distinct entities which act as summaries of pattern. They represent branching diagrams which depict patterns of shared, specialised characters (synapomorphies) found in a pair of taxa or sister groups. For the phylogenetic cladist, cladograms are abstract hypotheses of relationship which do not correspond exactly to the ancestral–descendant pattern of the unknown segment of the evolutionary tree that it may represent. For the transformed cladist, this is spurious, since the cladogram only depicts branching patterns.

In contract, evolutionary trees represent a summary of pattern with an added summary of process, the historical process of descent with modification that caused the pattern of characters. They can depict actual patterns of ancestral–descendant relations among taxa, rather than patterns of common ancestry alone. Finally, scenarios, as all cladists use the term, are not diagrams but represent historical narratives which attempt to describe which groups gave rise to which, as well as ecological changes and evolutionary forces which actually produced the adaptation which characterise the organisms discussion.

It is obvious that the notion of cladogram is slightly different in phylogenetic and transformed cladistic methods. Transformed cladists argue that cladograms depict only patterns, which *may* be consistent with a phylogenetic interpretation. Thus Nelson stipulates that '. . . a cladogram is an atemporal concept . . . a synapomorphy scheme.'[17] It is a summary of patterns of hierarchical character distributions in nature. For phylogenetic cladists, because a phylogenetic tree of species is both necessary and sufficient to follow the history of evolution on both the specific and supra-specific levels of biological organisation, a cladogram must depict patterns of taxa on the basis of descent (through the discovery of the appropriate uniquely derived characters and shared derived characters). In addition, cladograms are more general than trees because a cladogram is equivalent to, or summarises a set of evolutionary trees. In proceeding from left to right in the following argumentative chain:

cladogram   ⇒   evolutionary tree   ⇒   scenario

all cladists see an increase in information content with a complementary decrease in testability. Unfortunately, this final claim is misleading, because in proceeding from cladogram to scenario, there is really an increase in the range of tests that can be applied, such as the use of ecological tests in constructing scenarios.[18] Testability, as used by cladists, implies testability with respect to cladograms alone.

Another methodological shift concerns the distinction between species transformations and character transformations in cladistic analysis. Originally, both Hennig[19] and Brundin[20] regarded the species as the basic unit of cladistic analysis, species which were characterised by at least one unique derived character. Now, for a growing number of cladists (including transformed cladists), any monophyletic group which can be characterised by appropriate traits can function in cladistic analysis, regardless of whether that group is more or less inclusive than traditional species. In accordance with this shift, cladogram forks have undergone a change in representation from depicting speciation events to the emergence of evolutionary novelties. In early phylogenetic cladistic works, the nodes are taken to represent ancestral species, while present-day cladograms do away with nodes leaving the termini as representing speciation events and/or the emergence of unique derived characters (Fig. 28).

Fig. 28 The 'evolution' of cladograms.[21] (a) Hennig; (b) Nelson; (c) present day. Dots represent species and open circles are hypothetical common ancestors.

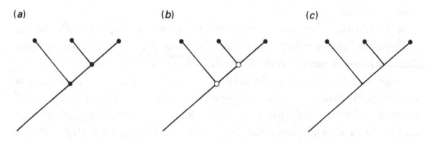

(a)   (b)   (c)

In effect the transformed cladists are attempting to construct classifications that are methodologically neutral with respect to evolution. Hence the distinctions between pattern and process, cladograms and trees, and species transformations. There are two further areas where transformed cladists have to uphold methodological neutrality with respect to evolution: the notion of synapomorphy, and its application in cladistic analysis. Nelson & Platnick specify that the concept of synapomorphy is definable as an element of pattern, a unit of resolution. Central to this notion is homology,[22] the relation which characterises monophyletic groups.[23] Synapomorphy has the same empirical basis as homology; both concepts are interdependent, and may be considered without reference to evolution, although an evolutionary interpretation may be applied to them without necessarily changing their empirical basis. Thus, a statement of synapomorphy implies comparison, and therefore homology, between certain characters of all other organisms, characters such as organs, more inclusive parts of other organisms, or even organisms as wholes. Both homology and synapomorphy are the same; in the sense of specifying a hypothesis of a set, homology implies generality, while synapomorphy implies relative, or restricted generality (that there is a subset included in . . .). Patterson takes this argument one stage further by specifying that symplesiomorphies are hypotheses of set, while synapomorphies are hypotheses of subsets within those sets. This is possible because a synapomorphy at one level *may* be a synapomorphy at the next higher level of set inclusion, while a synapomorphy at one level is always a symplesiomorphy at the more general level 'below'. This idea of levels of generality or universality can be expressed thus:

> symplesiomorphy and synapomorphy are . . . terms for homologies which stand in hierarchic relation to one another.[24]

Transformed cladists, in using the terms symplesiomorphy and synapomorphy, aim to drop connotations of 'primitive' and 'derived' (and necessarily any reference to the process of genealogical descent[25]) and replace them by specifying patterns of more or less inclusive sets, i.e. sets within sets. Thus, the character states represent either the general or less general condition.

There remains one final problem here: the determination of the hierarchy of nested sets in terms of universality.[26] Nelson & Platnick acknowledge that there is an implicit time element within the concept of character transformation. They suggest two techniques – comparative and ontogenetic.

(i) Comparative techniques

These include outgroup comparison,[27] which is an *indirect* technique: it depends not only on observations of character distributions or changes, but also on a hypothetical higher-level classification. To summarise, outgroup comparison involves the simultaneous analysis of many characters for deriving cladograms and interpreting various aspects of character evolution. Outgroup comparisons are used to study the most informative connection of the study world to the rest of the world, and this is specified in the outgroup rule which holds that two characters which are homologous and also found within the sister group, represent the general form, whereas the character found within the monophyletic group represents the less general character.

(ii)   Ontogenetic techniques

For Nelson & Platnick, this is a direct technique in that character changes are based on observation alone. The ontogenetic method is not based on Haeckel's law of recapitulation, where the sequence of developmental stages in ontogeny represents a vastly accelerated version of the sequence of adult forms in phylogeny, but in terms of Von Baer's law – generality.[28] Ontogeny passes from the simple to the complex, from the general (i.e. more widely distributed) to the particular.

For instance, in Nelson's example of flatfish eyes,[29] and his (1973) example of open visceral clefts versus closed visceral clefts,[30] open visceral clefts and an eye on each side of the head are more general than closed visceral clefts and both eyes on each side of the head, because animals showing the latter condition also show the former in early ontogeny. Of special importance in this approach is the possibility of stating Haeckel's biogenetic law in non-evolutionary terms, since 'general' and 'less-general' need not have an evolutionary interpretation. In the non-evolutionary sense, the hierarchy of synapomorphies represents the pattern of ontogenies of organisms. The crucial question here is whether the use of the ontogenetic method justifies a non-evolutionary interpretation and this will be discussed in detail in Chapter 6.

All transformed cladists acknowledge the superiority of the developmental approach over outgroup comparison. Ontogeny manifests an unfolding of pattern, which is not necessarily evolutionary. To the phylogenetic cladist, who argues that characters, like taxa, have particular times of origin and so arise at particular levels of phylogenetic descent, the methodological priority of ontogenetic analysis over outgroup comparison is not necessary.

To summarise, I would like to outline briefly the transformed cladists' view of classification construction. As we have seen, transformed cladists argue that natural taxa have unambiguous defining characters. All we have in classification construction are taxa and characters, and we should be searching for defining characters of natural taxa. The defining character, or suite of characters, for any natural taxon will be manifest in the ontogeny of all members of the group. Outgroup analysis has no special role except to polarise characters to match those determined by ontogeny. Homologies give us hypotheses about the relationships of taxa only in terms of nested sets. Homology can tell us nothing about genealogy. The patterns of characters defining various taxa can be expressed in a branching diagram, a cladogram; and to convert a cladogram – a set of phylogenetic trees – into one particular tree requires information beyond that which is available. With respect to classification construction the transformed cladists regard phylogenetic reconstruction as totally superfluous, so that the patterns of interested sets are directly converted into a classification. Classifications are little more than summaries of characters as evidenced in the hierarchical clustering of synapomorphies, and can be derived either from patterns of interested sets of synapomorphies uniting sister groups, or the cladogram (Fig. 29). Classifications do not represent the ordering of branching of sister groups but the order of emergence of unique characters defining natural groups, whether or not the development of these characters happens to coincide with speciation events.

Like the pheneticists before them, the transformed cladists emphasise the elimination of bias in classification construction. This is achieved through the use of a strict methodology. But implicit within such arguments is the assumption that such a methodology can justify a non-evolutionary interpretation. For example, Nelson's method of component analysis[31] states that cladograms are sets of evolutionary trees only one of which is topologically equivalent to the cladogram, and it is cladograms, divorced from evolution-

Fig. 29 Nested sets and cladograms.

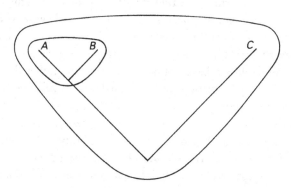

ary implications, which have the greater generality and testability, and which form the basis of systematic analysis. Nor do any evolutionary assumptions supposedly enter into the choice of cladogram which best summarises a set of data; *parsimony* or *maximal congruency* forms the basis for evaluating competing hypotheses. For three taxa, A, B and C, the number of characters in which A and B share a common state is compared with that possessed by C. The arrangement which produces the least number of contradictions and the greatest number of corroborations is the most congruent. Preference is for simpler hypotheses over complex ones.

Transformed cladists use the parsimony criterion in all cases where an alternative, evolutionary inference is rejected. Thus character transformations, as evidenced in ontogeny, are justified by reference to parsimony. As Patterson explains,

> . . . parsimony, as a criterion has nothing to do with opinions about evolution (whether it follows the shortest course or not) or any other process in biology. We accept the most parsimonious, or simplest explanation of the data because parsimony is the only criterion available. If parsimony is set aside, there can be no other reason than whim or idiosyncracy for preferring any one of a myriad possible explanations of a set of data.[32]

The ordering of organisms into groups need not have any biological significance; it is only a matter of discovering the simplest groupings.

From the foregoing, it should be clear that as in phenetics, transformed cladistics confronts problems in eliminating bias and rejecting explanation, so conforming to the descriptive attitude. However, justification for the elimination of theoretical presuppositions in transformed cladistics is different. Both pheneticists and transformed cladists are in accord over issues about the underdetermination of theory by data, and so have real doubts about the 'scenario' approach. But they differ in one important respect. The strong pheneticists are committed to rejecting the explanatory element from their classifications. In contrast, the transformed cladists have real doubts about the content of the underlying theory, and hold the revolutionary view that evolutionary theory is in a mess. For example, arguments over the tempo and mode of evolution have not been resolved, and rather than coming down in favour of punctuationism or gradualism, transformed cladists feel that theories of process are underdetermined by patterns. In distancing themselves from the applicability of evolutionary theory in classification construction, various rules of thumb, e.g. the irreversibility of evolution and uniform rates of evolution, are rejected. In transformed cladistics, the descriptive attitude is justified by calling into question the whole framework of evolutionary theory. Thus emphasis is placed on *internal*, rather than *external* issues (as in strong phenetics). All invocations to Popper are no more than a superficial 'dressing'.

Although the motivation in strong phenetics and transformed cladistics is different, both schools commit the same fallacy; because there are limits to the descriptive attitude, they cannot be practising what they preach. However, I will argue that phenetics is more consistent with the descriptive attitude than transformed cladistics, because the latter implicitly assumes either a Platonic world view or is unintelligible except in the light of evolutionary theory.

Transformed cladistics and methodological neutrality

In accordance with the descriptive attitude, transformed cladists incorporate a structural model which specifies a pattern of hierarchically nested sets. The construction of nested sets of synapomorphies is based on a developmental method which utilises ontogenetic data. Any incongruencies that exist between the various hypotheses of synapomorphy are eliminated using a parsimony method. Assumptions based on evolutionary processes are supposedly eliminated. At its most basic, transformed cladistics involves

$$\begin{array}{ccccc} \text{recognition of} & \Rightarrow & \text{nested sets of} & \Rightarrow & \text{classification} \\ \text{defining characters} & & \text{synapomorphies} & & \end{array}$$

However, this simple summary does not do justice to the range of interpretations that can be found in transformed cladistics. Much of what they believe in can be placed simultaneously in several of the categories that I described in Chapter 1; e.g. the strong and weak descriptive attitudes, as well as the anti-descriptive attitude typified in Zangerl's morphological method. No doubt, this is, in part, a failure of the transformed cladists to 'nail their colours to the mast'. They usually make out that they are practitioners of mainstream cladistics[33] (i.e. followers of the Hennigian approach), but their writings suggest otherwise.[34]

The varying interpretations that can be assigned to transformed cladistics hinge on the notion of methodological neutrality. Some critics[35] have suggested that transformed cladists advocate theory-neutral classifications; that is, classifications without theoretical implications. But this view is misguided, because transformed cladists do accept that classifications incorporate theoretical assumptions, since such implications are part of the motivation for constructing classifications. Indeed, they focus on how much theory is incorporated in classification construction, and, with respect to the finished product, just what theory (if any) is being appealed to. They suggest that cladistics is supposedly better methodologically because it does not rely on the supposedly unfalsifiable assumptions of evolutionary biology. And it is better empirically because it does not require as many assumptions that might not, after all, be true.[36]

While the results of transformed cladistics may be at odds with some theories, their method is without theoretical prejudgement. Clearly, this desire for methodological neutrality implies that all bias can be removed in classification construction.

In adopting methodological neutrality, two possible versions can be set up in transformed cladistics. In the first, methodological neutrality can be invoked to eliminate bias with respect to the aims, goals and purposes of classification. In particular, the role of evolutionary theory in classification is rejected because evolutionary theory is in a mess, and the best strategy is to distance oneself from it as far as possible. Hence, all evolutionary assumptions are rejected in classification construction. This version is clearly consistent with weak descriptivism since a minimal number of theoretical assumptions are incorporated in classification construction (provided that they are not evolutionary). Such a strategy also avoids problems of underdetermination of the theory by data.

It is in this sense that transformed cladists argue that the principles of cladistic analysis do not concern evolution, but involve the general goals of biological classification and the means by which they are realised. In the context of cladograms and characters, Nelson & Platnick state that

> By 'synapomorphy' we mean 'defining character' of an inclusive taxon. True, all defining characters, in the phyletic context may be assumed to be evolutionary novelties. But making this assumption does not render it automatically true; nor does it change the characters, the observations on which the characters are based, or the structure of the branching diagram that expresses the general sense of the characters i.e. that there exist certain inclusive taxa . . . that have defining characters.[37]

The point being, that if the cladogram is made necessary by the characters and the characters by observation, then there seems very little in the way of theoretical assumption here. Thus classification construction is methodologically prior to phylogenetic reconstruction.

In the second, stronger version emphasis is placed on the rejection of explanation, as expressed in the decoupling of pattern and process. All causal input is rejected. As Platnick notes

> . . . one needs no causal theory to observe that of all the millions of species in the world today, only about 35,000 of them have abdominal spinnerets. One also needs no causal theory to observe that of all the

millions of species of organisms in the world, only about 35,000 of them have males with pedipalps modified for sperm transfer. One needs no causal theory to observe that the 35,000 species with abdominal spinnerets are precisely those 35,000 with modified male pedipalps . . . [a]nd one needs no causal theory to observe that such massive congruence of characters leaves no reasonable doubt that spiders constitute a real group that exists in the present day world.[38]

This view reflects the strong version of the descriptive attitude,[39] because over and above the attempt to produce patterns which are derived from various rules that are methodologically neutral with respect to evolutionary theory, transformed cladists also hint at recurrent patterns which suggest an alternative explanation. Indeed, the search for recurrent patterns is indicative of a general approach since explanatory appeal is not necessarily made to evolutionary theory (or any other theory). This general approach does not presuppose recourse to hidden causes. It is important to note that the justification for the suspension of hidden causes is based on the rejection of a particular theory, i.e. evolution, in the methods of classification construction, and the subsequent extrapolation to the rejection of all hidden causes. In this strong version, then, the transformed cladists argue that their resulting classifications question the status of evolutionary theory itself.

The final interpretation that can be assigned to transformed cladistics does not directly involve methodological neutrality, and is based on the idea that

A cladogram is a summary of pattern, the pattern of character distribution, or of hierarchy in nature – what pre-Darwinians called 'the natural hierarchy'.[40]

This pattern is a reflection of the *one* natural order, i.e. 'that nature is ordered in a certain specifiable pattern',[41] and is seen in

the unfolding of pattern, increase in complexity, or differentiation . . . in the development of individual organisms – the process of embryology rather than of descent with modification.[42]

With this emphasis on the *one* order, the methods of transformed cladistics might represent a return to pre-Darwinian methods of classification. For example, there is a superficial similarity with idealistic morphology, as typified in Owen. Owen specified the notion of an underlying pattern, documented by a seeming unity of plan in certain organisms, i.e. a persistence of structure or order. In apparently following this approach, transformed cladists are realists about underlying patterns. Consequently, their attitude could be anti-descriptive, since idealistic morphology, in the pre-evolutionary period, incorporated an underlying, explanatory theory.

Despite these interpretations, it is still possible to characterise a basic world view[43] in transformed cladistics. This world view can be characterised as follows.

(1) Nature is ordered in a single specifiable pattern, and that order is hierarchical and can be detected by sampling characters.

(2) Classifications serve as hypotheses about the hierarchical distribution of characters.

(3) Only natural groups may be unambiguously defined, and the defining characters can be found by direct inspection of the ontogenies of component members of the group which yield replicated, interested sets of synapomorphies (defining characters).

(4) Our knowledge of evolutionary history, like our classification, is derived from the hierarchical pattern thus hypothesised.

In addition, transformed cladists reject biology based on levels of organisation and mechanistic evolutionary principles, and espouse a world view based on studies of uniformity and diversity of observed traits. As a science of pattern, transformed cladistics holds out the possibility of a reconstruction of the natural order that is not dependent on evolutionary assumptions, and which utilises the comparative approach. This approach deals with three elements:

(1) similarities and differences in the attributes of organisms, (2) the history of organisms in time, and (3) the history of organisms in space.[44]

which are summarised in Croizat's maxim:

Space, Time, Form: The Biological Synthesis.[45]

In effect, the comparative approach searches for recurrent patterns.

Because of the various interpretations that can be accorded to this world view, there is a need to clarify not only how transformed cladistics fits into the descriptive attitude, but also to explicate the standards by which it judges itself. The central point to be grasped in all three interpretations is whether the claim for methodological neutrality can support the weak version alone, or whether the strong version is also a natural corollary of it. Those critics who advocated a theoretically neutral interpretation of the results of transformed cladistics failed to question the weak version in the first place; they assumed that methodological neutrality in classification construction supported theoretical neutrality of the results to live issues in evolutionary theory. This is obviously misguided since we must first examine how sensitive the 'neutral' assumptions in transformed cladistic methodology are to evolutionary theory. Thus, in the weak version, we should examine the relevance of evolutionary theory in classification construction, and then, on the basis of this, question the claims for neutral paterns with respect to evolutionary theory.

These problems can be summarised in the following questions.

(1) Do the transformed cladists have a theoretically neutral methodology in classification construction?

(2) Does this neutral methodology justify a stance of theoretical neutrality in the finished product, *vis.-à-vis.* evolutionary theory?

(3) Does the rejection of evolutionary explanations substantiate a claim for the rejection of all hidden causes?

Questions (2) and (3) have important implications for the notion of pattern in transformed cladistics. The former focuses on the elimination of those patterns which are taken to constitute evidence for evolutionary theory, i.e. do these patterns necessarily presuppose modification by descent? If significantly different patterns were to be found, then this evidential basis would be in doubt. Two courses of action could be possible at this juncture: either a replacement theoretical framework is invoked (in accordance with the third interpretation based on idealistic morphology), or a case is made for calling these patterns 'general' in that there is no necessary recourse to hidden causes (question (3)). Although the transformed cladists would argue for the latter course, it is not clear that they would succeed in abandoning a theoretical framework. For, while the discovery of odd recurrent patterns would certainly provide a strong justification for questioning the status of evolutionary theory, it is not apparent whether such discoveries would justify the suspension of belief in all hidden causes. In other words, how closely does transformed cladistics conform to the strong version of the descriptive attitude?

In the rest of the chapter, I will examine this problem through a comparison of phenetics and transformed cladistics. Such a comparison is necessary before any critical assessment can be made of transformed cladistics on its own terms. I will be dealing primarily with question (3), and will assume, for the moment, that the transformed cladists do have a neutral methodology with respect to evolutionary theory. This assumption will be examined in the next chapter.

Phenetics and transformed cladistics: how similar are they?

A cursory glance through Nelson & Platnick's *Systematics and Biogeography* reveals an indiscriminate use of phenograms dressed up as cladograms,[46] implying that the techniques of phenetics and transformed cladistics are in the same business, despite claims to the contrary by the latter. As we have seen, both Farris[47] and Mickevich[48] argue that cladograms may be used as a basis for phenetic classifications because they give a greater measure of stability, thereby optimising Gilmour naturalness. However, I also noted that the standards for comparison were inconsistent because the cladists interpreted naturalness in terms of monothetic groups, when, in reality, the pheneticists used polythetic groups. In the light of this deficiency, it would seem sensible to make a comparison between phenetics and transformed cladistics on the basis of monothetic methods of grouping alone. Of course, transformed cladists would reject this, on the grounds that phenetics deals primarily with polythetic methods which are supposedly inferior to cladistic methods. While there may be some justification for this argument, a comparison of monothetic methods within phenetics and transformed cladistics does shed some light on the different degrees to which strong descriptivism is adopted by each school.

In phenetics several monothetic methods have been suggested. Lockhart & Hartmann[49] suggested a clustering method in which the groups, although polythetic, are made monothetic by discarding all characters that vary within them. In general, the aim is to select a character which bisects the set of individuals at each stage. Characters are chosen which minimise information loss with respect to other characters, or to some other related function, e.g. Lance & Williams[50] used functions based on X^2. In specifying characters with maximum predictive value, monothetic methods are advantageous because keys and hierarchies are readily made. They also produce definable groups which are of great use in taxonomy, and it has been claimed by Lance & Williams[51] that monothetic methods distort the data little more than polythetic methods. Despite these advantages, pheneticists use monothetic methods infrequently. They argue that there is always the danger of serious miscalculation in constructing natural phenetic groups, since any defined feature involved in division could be aberrant. Thus monothetic groups 'do not yield natural taxa except by a lucky choice of the feature used for division.'[52] Jardine & Sibson[53] have also noted that monothetic methods involve an arbitrary choice of procedure because they require partitioning into *two* subsets at each stage. These deficiencies emphasise the point that such methods produce a hierarchical classification by choosing a dignostic key based on the available characters.

The pheneticists further argue that polythetic groups have a higher information content than monothetic groups, a point which the transformed cladists deny. For the transformed cladist the all important task is to search for diagnostic characters which are then used to define groups of organisms. Such groups are said to be real in the sense that they are definable. Since such groupings are, by definition, specified by at least one character, then the information content is supposedly higher. Indeed, the transformed cladists argue that while the pheneticists can only measure information content by counting groups, they can base their own measurements on the number of groups, and what these groups contain, since each group is defined by at least one character. As it stands, this argument over information content is misguided.

Information content, as employed by transformed cladists, is derived from the observation that for any particular classification, there is a corresponding cladogram that specifies the information content of the classification. Information content is expressed in two forms: *component* (C) and *term* (T) information.[54] *Components* of a cladogram are branch points (nodes) defined by the terminal taxa to which they lead. Terminal taxa from each node comprise Nelson's *terms*, whose information is the individual diagnosis of sets of characters for each taxon. The information content of all components is the number of components for a given cladogram expressed as a fraction of the total number possible, the total being two less than the number of terminal taxa. The information content of all terms is the sum of all information for all components. Of course, in setting up this expression of information content, the classification does not specify in what context the components should be interpreted (phyletic, phenetic), so that the information content of the classification is comprised only of characters (general statements of synapomorphy) and the groups they define.

The transformed cladists make the implicit point that the use of monothetic groupings gives higher information content than the polythetic groupings of the pheneticists. But they fail to point out that if the pheneticists were to use monothetic groups, then the information content would be the same for each school. The pheneticist would acquire further information concerning the defining characters of each taxon. However, the fact that the pheneticist rarely, if ever, resorts to the inclusion of monothetic groups is important in itself. For him, the maximisation of information content is not the sole aim of classification. Classifications, if they are to function as devices for the storage and retrieval of information, should, above all, be stable. In the context of the descriptive attitude, this point is crucial, because while information content is an important factor, it must at all times, be offset by stability. Thus, arguments to see whether phenetics or transformed cladistics conform better to the descriptive attitude in terms of maximising information, should be considered in the light of other aims of classification, such as stability and internal coherence. For both schools, the question of stability is closely tied up with ideas concerning congruence and reality. It is to these issues that I now turn.

The transformed cladists give an epistemological justification for adopting the descriptive attitude. This can be seen in Nelson & Platnick's[55] discussion of the orderliness of nature. Their views are based on Hennig's original formulation of the task of systematics;

> . . . the investigation and presentation of all relations that exist among living natural objects. It is not to introduce order into the manifold of particular phenomena, but to represent this order.[56]

In reiterating and expanding on Hennig's realism, Nelson & Platnick claim that

> [t]here is no direct way for us to actually determine whether the disorder that we perceive exists in nature or only in our own hypotheses. The purpose of science, however, is to explain regularities in nature, and if we give up the search for regularities, we also give up the game of science. So we may assume that the order really does exist, and that we, and not nature, are responsible for the apparent disorder.[57]

The aim is to see order in apparent chaos, through critical observation and hypothesis testing. In effect, the transformed cladists are claiming that they can document an objective reality set apart from causal factors in a theory. The patterns that they detect in nature are real.

In contrast, pheneticists do not assume directly that any apparent disorder is necessarily a function of the classifier. They are, as we shall see, forced to take a more instrumentalist attitude by arguing that their classifications are tools for serving a specific function, and do not directly concern the representation of an objective pattern in nature. For the pheneticist, because the method used in classification directly influences our perceived order, the question of detecting orderliness in nature is not so much an epistemological issue, as an empirical one. This is reflected in their rejection of early studies on congruency.

Such studies were based on the hypothesis of non-specificity, which assumed that there were

> no distinct large *classes* of genes affecting exclusively one class of characters such as morphological, physiological, or ethological, or affecting special regions of the organism such as head, skeleton, leaves.[58]

From this hypothesis it followed that there were no *a priori* grounds for favouring one character over another, so that congruency was a question of the number of characters required to obtain reliable results, i.e. an estimate of the number of characters required gave an estimate of congruency. In testing this measure of congruency, Sokal & Sneath[59] suggested two hypotheses: (i) hypothesis of factor asymptote and (ii) hypothesis of matches asymptote. In the former, emphasis is placed on information obtained about a single organism or OTU by studying its characters, such that a random sample of characters should represent a random sample of the genome of the OTU. With the inclusion of more and more characters, there is an increase in information with each character. This information increase is large at first, but decreases, and after a certain number of characters have been recorded, will approach close enough to an asymptotic value to make the further inclusion of characters unnecessary. Stability is therefore maximised by congruency between different sets of characters.

In the hypothesis of matches asymptote, it is assumed that as the number of characters increases, the value of the similarity coefficient becomes more stable. No additional characters are required because an objective measure of similarity has been found. Such an objective estimate might be based on the nucleotide sequences of the DNA of the OTUs. In this respect, Sokal & Sneath even talk of estimating objective similarities between fundamental genetic units. They are realists about patterns.

But, the idea that congruency of characters is an indicator of objective patterns in nature is false. Incongruencies between classifications based on different organs or different life-stage histories were found, implying that the hypotheses of asymptotes were wrong. Identical classifications were not produced from different sets of characters for the same OTUs. The pheneticists were forced to take a more instrumentalist attitude because of this intrinsic incongruence with increasing numbers of characters. Rather than measuring stability through character congruency within a given method, emphasis was placed on the use of different methods to give similar arrangements. Congruency came to be defined as

> the degree of correspondence beween arrangements of OTU's in a classifiction,[60]

implying that the more congruent the resulting classifications, then the more stable the methods. Classifications were objective in the context of stable methods.

For the pheneticist, the stability of a classification is not just a question of whether nature is orderly or chaotic; it is also an empirical problem. While the world may be orderly (as typified in their early idea of measuring similarities between organisms to obtain objective, i.e. real, hierarchies), the discovery of such an order is dependent upon the methods of classification. Nevertheless, there is a close similarity between the early phenetic views of objective measures of similarity and the transformed cladists' discussion of the one correct order. However, the pheneticists have had to weaken their claims, while the transformed cladists have remained firm realists. This is because, in transformed cladistics, the question of congruency and reality is totally determined by the method. With a method that searches for defining characters, all perceived incongruency must be eliminated, so that the stability of the resultant classification is a measure of objectivity in terms of the real order. For example, in a three taxon statement, with taxa X, Y and Z, all but one arrangement is rejected (falsified). While all the arrangements may have a degree of incongruency between characters, it is the least incongruent arrangement which is accepted (i.e. all but one hypothesis is eliminated) because it conforms more closely to the real order.

It is over the status of perceived incongruencies in patterns, that transformed cladists equate controversies in taxonomy in terms of 'artificial' and 'natural' systems, involving a choice between the acceptance or rejection of incongruence as a feature of the real world. They discuss this on the basis of one's view of the nature of the scientific pursuit, and not in terms of empirical issues. Thus, if nature is inherently chaotic then all systems will be artificial; if nature is ordered then there is the one and only natural system. For transformed cladists the major difference with phenetics is that

> whereas phenetic methods are willing to accept incongruence between characters as a feature of the real world, cladistic methods regard the discovery of apparent incongruence as an indication that the taxonomist has made a mistake.[61]

Platnick & Nelson do admit, though, that phenetics includes real (but random) aspects of nature,

> and in its favour it must be admitted that there is no way for us to actually determine whether the disorder that we perceive exists only in nature, or only in our own hypotheses.[62]

For the transformed cladist the question of congruency and reality is a philosophical question, not an empirical one. It is implicitly assumed that the method can uncover the natural order, implying that incongruence can in principle be eliminated with complete knowledge. But, because transformed cladists are less concerned with an empirical order, and focus on questions of methodology, then their classifications are less indicative of the real order. Just as the transformed cladists do now, the pheneticists used to specify congruency in terms of numbers of characters. They were realists about patterns. But, having discovered that congruency between characters does not necessarily give the real order, they redefined congruency in terms of different methods producing similar arrangements. For their part, the transformed cladists have ignored this empirical evidence against congruency between characters, because they are committed to one method, which involves the search for defining characters. Congruency, for the transformed cladist, can only be defined in terms of characters.

There is obviously an element of naivety in the claim that congruency implies a non-artificial order, because no account is taken of the possibility of either an incongruency actually existing in the patterns when they appear congruent, or of a supposedly incongruent pattern which is actually congruent. Congruency does not necessarily uncover the natural order. Furthermore, the notion of congruency in transformed cladistics hints at a real order hiding behind the 'appearances'. While the 'appearances' could be misleading, there is still an underlying natural order, 'whereby each cladogram tells the same, or part of the same story,'[63] and which can be uncovered by eliminating incongruence. In this way the hypothesis of perfect congruence specifies such an order. Indeed, congruence and reality are one and the same thing; varying degrees of congruence specify different degrees of orderliness (and reality). However, this notion of an underlying order implies that the transformed cladists cannot be conforming to the descriptive attitude in the way that the pheneticists do, because they do not pay enough attention to empirical considerations.

Given these observations concerning reality and congruence, we are now in a position to clarify the question of stability in classification. It should be evident that pheneticists equate the stability of classification with repeatability of results, in the sense that the most stable classification is the one which gives the most congruent results on the basis of different methods. Furthermore, as mentioned in Chapter 4, stability is measured through predictive power and information content. I also noted that the strong pheneticists inadvertently incorporated a redundant notion of information content, implying that stability and predictive power are one and the same thing. It is interesting to note that the transformed cladists recognise this latter point:

> In reality, stability and predictive value are not in conflict with each other, they are one and the same thing. The most stable classification is the one that most successfully predicts the structure of newly acquired data, and those classifications that are not successful in predicting new data will have no stability. They will be discarded in favour of more successful classifications – classifications that have higher predictive value, and, hence in the long run, greater stability.[64]

But, for them, stability is also predicated on a real order concealed by 'noise', not on the repeatability of methods giving rise to similar solutions. In accordance with this, the search for character congruence is based on predicting the structure of newly acquired data. Thus

> To the extent that there is order in nature, to the extent that existing
> classification is an accurate hypothesis about that order, and to the
> extent that our hypotheses about properties and their distribution
> among organisms are correct, such predictions will be successful.[65]

In other words, there is the one correct cladogram that can be tested by the
prediction that synapomorphies will be congruent with one another. So, the
transformed cladists adopt a notion of stability which assumes that
congruence is a true indicator of the order in nature. But, because of their
commitment to the one order, it can only be assumed that incongruency
specifies a superficial order (an 'appearance'). This is clearly antagonistic to
the descriptive attitude where emphasis is placed on empirical considerations, and not just methodological rules. Of course, this emphasis on methodology is, in part, a justification for arguing that the search for defining
characters of groups enhances stability. But, as the pheneticists noted,
monothetic methods do not necessarily give stable classifications, because
the defining characters are often wrong. In the light of this, rather than
rejecting or modifying their methods, the transformed cladists have argued
that for something to be real, it has to be definable. It follows that definable
groups are the only real, stable groups. Yet, the use of definability as a criterion of group reality is bizarre, and again forces the transformed cladists to
assume the existence of a hidden order. But this is the price they pay for
ensuring stability of their classifications while retaining monothetic methods.
In expanding upon this point, it is necessary to compare the basis for character choice between phenetics and transformed cladistics.

For the transformed cladist, the only characters of interest are those that
are defining characters of groups. Such defining characters are synapomorphies
which specify an inclusive taxon. No reference is made to character states
because

> The concept of a character state is potentially misleading. To view
> some character X as being composed of [say] three states implies that
> the characters are alternatives when they are actually additions.[66]

All characters are seen as modifications of other, more general characters.
Thus, in vertebrates, fins and limbs are normally regarded as alternative
states of a character (paired pectoral and pelvic appendages), while for transformed cladistics, limbs are really a modification of fins. In contrast,
pheneticists use the notion of character states, and derive a measure of
overall similarity (though subsequent character selection may be made on
the basis of initial group recognition). For them, all characters and their
respective states are relevant, irrespective of whether they are taken to be
'general' or 'less general' within a transformed cladistic context. This use of

general and less general (or absence/presence of) characters by transformed cladists, implies a form of character weighting.

The concept of synapomorphy as restricted generality presupposes continuity, or modification, from one character to another. In contrast, characters and character states do not and, on the basis of what can be 'observed', cannot incorporate less bias. The phenetic use of characters is more consistent with the descriptive attitude. More importantly, from the standpoint of information content and stability, the choice of synapomorphies for specifying groups represents only a subset of all possible characters at any given level. The aim is to give greater stability with the incorporation of fewer characters. In phenetics, the aim is to maximise the number of statements that may be made about all character states, not just synapomorphies, because stability is better achieved with all characters. This difference in attitude is again a function of the search for either monothetic or polythetic groupings.

Now, on the question of definition and recognition of natural groups in nature, the transformed cladists argue that

> monophyletic groups, like species are real: they exist outside the
> human mind, and may be discovered. Homology, which exists in the
> human mind, is the relation through which we discover them. Non-
> monophyletic groups, unlike species, have no existence in nature. They
> have to be invented, are not recognised by homology, and exist only in
> the human mind.[67]

Typical of such non-monophyletic groups are paraphyletic groups which are characterised by symplesiomorphies. Because such groups lack defining characters, they are uncharacterisable and therefore exist only in the mind of the taxonomist.

As it stands, transformed cladists such as Patterson appear to be making an ontological claim over the existence of groups. Those groups that are definable, exist in nature. Those that are not, do not exist. But to use definability as a criterion of reality is misguided. On the one hand, it implies that groups are unreal if they are based on many incongruent definitions but at the same time show congruence of similarity. Even in the extreme cases where no defining characters are present, it does not follow that such groups are unreal. For example, Wittgenstein's family resemblance groups were undefinable, but nevertheless existed. On the other hand, it is equally possible that defining characters can be found in the absence of congruency of similarity. In this case, there appears no reason to suppose that such a group is real. As if in anticipation of this problem, Nelson & Platnick state that it is helpful to distinguish between groups that really seem to exist from 'groups' that may be defined. That a grouping may be defined by one or more 'characters' does not mean, therefore, that the group has any existence in the real world.[68]

Unlike Patterson, Nelson & Platnick decouple reality from definition, implying that it is possible to define paraphyletic groups without specifying their existence. But how then, are we to specify a real group, if it is not by definition alone? In the absence of any causal factor for reinforcing the notion of reality,[69] it is hard to envisage why one group is real while another is not, solely on the basis of the transformed cladists' notion of character congruency. As we have seen, an appeal to congruency of characters is not sufficient to justify the objective existence of a group, implying that Nelson & Platnick's argument still rests on a connection between definition and reality.

It certainly is not clear why groups defined by synapomorphies should be real, while all other possible groups are not, purely on the grounds that they are either defined differently or are undefinable. Yet some undefinable groups, e.g. paraphyletic groups, still exhibit predictive properties. The transformed cladists must provide some justification for focusing on defining characters, other than through reality. Under the Hennigian system, the interest in definability is more understandable because it is implicitly tied up with the assumption that evolution is irreversible. But in transformed cladistics there seems little justification for the non-existence of paraphyletic groups simply because they do not express synapomorphies. In the absence of any causal factor, the transformed cladists fail to substantiate claims for 'existence' and 'reality' on the basis of definition alone. This failure stems from a confusion of the question of congruence and reality with the question of congruence and objectivity. The pheneticists attempt to document an objective order through measures of congruency: different people using different methods obtain similar results. The most objective order is the most stable one. In contrast, the transformed cladists are not interested in congruency and objectivity; they only pay lip service to the question of congruency by assuming that the notions of reality and objectivity are one and the same. Since defining characters specify real groups, then the real order is objective and stable. But this stress on defining characters does not specify an objective order; with the possibility of mistaken defining characters, the transformed cladists can, at best, only give diagnostic criteria for group recognition. They are attempting to diagnose the one true order. This order cannot be objective in an empirical sense, implying that stability is not necessarily a matter for empirical investigation because it can be justified by appealing to a hidden order.

Given that the transformed cladists do not specify the notion of an objective order in the sense that the pheneticists do, in what way do they characterise this order? If the order is not objective, then what is it? From the discussion so far, several points should be evident: firstly, there is an underlying order which is not evolutionary, or even causal in any sense. Secondly, because this underlying order is uncovered through the elimination of incongruency, which is rarely, if ever, achieved, then the order that we obtain in practice is only an imperfect reflection of the order 'below'. There is an ideal order which is concealed from us. Thirdly, the insistence on a proper definition reinforces the notion of an underlying order. Since reality is linked to definability, then this order is real. Interestingly, there is much in common between these attitudes and Platonism.

Platonism and transformed cladistics

In suggesting that transformed cladistics closely conforms to a Platonic world view, it is important to distinguish between several philosophical positions which are encompassed by the term 'Platonism'. In the first instance there is the doctrine in Plato, of the existence of Forms, of really existing abstract entities of which physical objects are imperfect copies. Secondly, Platonism also implies a position holding that something can have an essential property in virtue of its definition, or as described in a certain way. Finally, Platonism, in a form exemplified by Aristotle, maintains that some objects – no matter how described – have essences;[70] they have, essentially or necessarily, certain properties, without which they could not exist or be the things they are. As far as transformed cladistics conforms to a Platonic world view, it is only in the sense specified by the theory of forms that this occurs and *not* in the other senses.

According to the traditional interpretation of Plato's philosophy, universals such as man, goodness and triangularity existed in a realm of appearances. The name 'Fido' referred to an individual dog, while the word 'dog' referred to the Idea or form dog in the separate, eternal and unchanging realm of Ideas. Individual dogs might come and go but the Idea of dog was immutable. Thus, these eternal transcendent realities directly apprehended by thought were contrasted with the transient, contingent phenomena of our empirical existence.

Plato's theory of Forms refers only to objects of knowledge,[71] that is, existing things. The objects of knowledge or Ideas existing in an eternal realm which are not propositions but things: objects independent of the mind. These objects can only be described by definition, implying that there must be a reality that the definition is true in virtue of. For example, the reality described by the true definition of justice is not one located in any particular time or place, or one that changes, but is, in every respect (is fully) just. In other words, what is taken to exist is also what is taken to be true. This view implies that for any word to have meaning, there should be some entity to which it stands in the same relation as the name 'Fido' does to the dog.

Plato's theory of Forms has long been regarded as untenable, not least for coupling definition and reality together. In this respect, it is a mistake to suppose that for words to have meanings is for there to be entities for which they stand. Furthermore, in specifying a metaphysical order, it is not obvious that the observable is real in any sense, implying that while we can only base our metaphysical order on observables, this empirical basis is undermined because it is in no sense real. Nevertheless, the similarity between trans- formed cladistics and Plato's theory of forms is striking. Both search for a real, eternal, non-causal order which exists independently of the realm of 'appearances', but which can in principle be specified. Both assume that for something to exist, it must be true in virtue of its definition. Given this simi- larity, it is not surprising that the transformed cladists could be interpreted as advocating a version of idealistic morphology. But this interpretation misses the point, despite similarities in the search for a limited number of fixed and unchanging forms which reflect the variable world of phenomena. The dif- ference lies in the question of causality. Plato's theory of forms has no causal input, and it was for this reason that Aristotle criticised it for failing to be an explanatory device. Furthermore, the typological view implicit in pre-Dar- winian idealistic morphology assumed the existence of essences: fixed and timeless possibilities of existence, in which the definition of a thing expressed the thing's essence, the characterisation it had to have to be the thing it was. Such essences had explanatory value. In transformed cladistics there are no such essences. Instead, there is some notion of an ideal order which is concealed from us, and which is only partly reflected in the order that we discover in nature. Given these differences, it is clear that trans- formed cladistics is not anti-descriptive.

The fact that the transformed cladistic notions of reality and order are only comprehensible in the light of certain Platonic assumptions has import- ant ramifications for situating transformed cladistics in the strong version of the descriptive attitude. Despite the superficial similarities[72] between trans- formed cladistics and phenetics (i.e. in phenetics the phenogram is made necessary by the characters, and the characters by observation, while corres- pondingly, for the transformed cladist, the cladogram is made necessary by the characters, and the characters by observation) there are major differences in the perception of patterns and in the formation of characters which lead to different standards in the comparison of classifications. While the pheneticists are committed to a rejection of the explanatory element, the transformed cladists are committed to the rejection of evolutionary patterns, which they take to be equivalent to the rejection of all hidden causes. However, it is only in a weak sense that the transformed cladists reject hidden causes. Although they make no reference to a causal theory, they still appeal to a world view that has no implicit notion of causality. The very fact that some world view has been appealed to implies that transformed cladistics does not conform to the strong version of the descriptive attitude as closely as phenetics does.

This failure is due to the transformed cladists' obsession with methodological issues. The strong pheneticists advocate the rejection of explanation on misguided philosophical grounds. The transformed cladists use methodological issues to back up their rejection of evolutionary theory. It is this appeal to methodology that inadvertently leads to a Platonic world view. Similarly, the transformed cladists also believe that Popperian methodology is crucial to their enterprise. This, too, is misguided. Like the pheneticists before them, the transformed cladists misrepresent philosophical attitudes, and in the remainder of this chapter I want to unpack the disanalogy between transformed cladistic methodology and Popperian methodology.

Popperian perspectives

Kuhn[73] remarked that when scientists fall out, they frequently resort to philosophy both to justify their own position and to attack those of their opponents. The transformed cladists (and phylogenetic cladists for that matter) are no exception; their espousal of Popper's special brand of hypothetico-deductivism is exclusively aimed at ruling out certain types of systematic procedure,[74] since it calls for realism, the search for truth (as opposed to consensus) and openness to criticism and testing.[75] According to the transformed cladists, both pheneticists and evolutionary systematists fail to uphold these criteria and so their methods must be rejected. Transformed cladists justify this attitude on the grounds that the philosophy of science has a normative role in taxonomic disputes. According to Platnick & Gaffney

> This . . . view . . . seems to be supported by some of the controversies that have erupted over the theory of systematics during the last two decades, many of which have been largely contingent on philosophical questions.[76]

There are two problems with this view. Firstly, most invocations and references[77] to Popper are largely a 'red-herring' because the transformed cladists do not follow Popper's methodology. Secondly, those aspects of Popper's philosophy which are more in line with transformed cladistic thinking, such as the search for truth and the plea for realism, have led to inconsistencies which can only be salvaged within the framework of a Platonic world view.

Transformed cladists' enthusiasm for Popper, no doubt, has some part in their rejection of evolutionary theory, and thus in their rejection of evolutionary classifications. If one emphasises falsifiability, then a fairly common complaint of evolutionary theory is that it is not falsifiable.[78] Indeed, Popper originally argued that there were no laws of evolution:

> The evolution of life on earth, or of human society, is a unique historical process.[79]

Darwinism is, therefore, not a testable scientific theory, but a metaphysical research program, a possible framework for testable scientific theories.[80] But this line of argument cannot be taken as motivation for the emergence of transformed cladistics, since their motivation is based on internal issues: evolutionary theory is in a mess, and this *alone* provides the justification for the shift away from using evolutionary assumptions. Questions of falsification are irrelevant and the incorporation of Popper's philosophy is at best only a *post hoc* justification.

More importantly, supposed instances in transformed cladistics where a Popperian methodology has been followed are rare. Transformed cladists, in following Popper, aim to construct classifications according to rules which allow one to put in, and retrieve, a large number of potentially falsifiable statements about the objects being classified. They argue that, typically, in a three-taxon statement, with taxa X, Y and Z, characters are used to falsify all but one arrangement, so that the least rejected hypothesis for the three-taxon statement is accepted. But this is certainly not Popperian, where theory growth is characterised in terms of problem solving:

$$P_1 \quad \Rightarrow \quad TS \quad \Rightarrow \quad EE \quad \Rightarrow \quad P_2 \text{ [81]}$$

(where P – problem; TS – tentative solution; EE – error elimination). Thus, a particular hypothesis, H_1, is modified in the light of possible falsification, and amended to H_2. Popper is advocating the expression of

> hypotheses in a form in which an attempt at falsification can be made most easily; not a choice amongst hypotheses (cladograms) *all* of which have already been falsified at least once, but a choice among new hypotheses none of which has yet been tested.[82]

There is clearly a disanalogy with Popper in the method of discovering the one true classification. Transformed cladistic methodology is more like a keying procedure – such as would be used in the identification of chemicals, rather than in Popperian science. There is too little theoretical input for it to be Popperian (since Popper also calls for theoretical depth).

Admittedly, Patterson[83] has argued that falsification should not be all that is taken from Popper; rather, Popper's insistence on the critical method and on the possibility of progress through one's own ideas and those of others should be emphasised. Patterson believes that we should strive to express our conjectures (of homology, of monophyletic groups) as clearly as possible, so that they are accessible to criticism. Yet, this really is nothing new, and should be part of any sound scientific enterprise. It certainly does not necessarily imply that transformed cladistic methodology is Popperian. If anything, the transformed cladists' devotion to Popper has, in practice, focused on questions of truth and realism, rather than hypothesis testing. Questions of truth and realism are, for Popper, inextricably tied up with the question of theories and scientific explanation. The transformed cladists, for their part, reject this insistence on explanation, but retain notions of truth and reality, thereby forcing themselves into a Platonic framework. As with the strong pheneticists, appeals to external issues have only been characterised by a high degree of disanalogy with the approved philosophical perspective, as well as internal inconsistency.

6

Transformed cladistics and evolution

In this final chapter my aim will be to examine whether transformed cladistics succeeds on its own terms. That is, do the transformed cladists uphold a theoretically neutral methodology (with respect to evolutionary assumptions) in classification construction? If it can be shown that homologies define hierarchies of groups in nature in the absence of evolutionary assumptions, then the resultant claim that these hierarchies are evidence of causal necessity (which would imply that they are natural) could also be upheld. I will therefore be examining questions (1) and (2) (p. 145) with a view to arguing that if the transformed cladists wish to avoid a Platonic world view, then their methods are unintelligible except in the light of evolutionary theory.

In order to show that there are no evolutionary assumptions in their methodology, transformed cladists must substantiate the claim that groups exist regardless of whether some causal theory is needed to explain their existence. For the transformed cladist this existence is guaranteed by congruence of homologies, such that homologies may be hypothesised and tested by a rational procedure that has no necessary dependence on evolution. Two elements must be teased apart in this methdology. First, what constitutes the structural model and, second, the methods for realising such a model. The model assumes that features shared by organisms (homologies) manifest a hierarchical pattern in nature,[1] while the method stipulates that homologies are resolved by Von Baer's law (p. 138), and that ingroup and outgroup analysis have no special role except to polarise characters to match those obtained by ontogeny. Neither the model nor the method should incorporate any evolutionary assumptions, and my discussion of the internal consistency of transformed cladistics will focus on these two aspects.

On a more general note, if the transformed cladists do succeed in purging all evolutionary assumptions from their methodology, then they still have to face the charge of resorting to a Platonic world view. No doubt the transformed cladist would attempt to avoid a Platonic interpretation because it undercuts some of his fundamental assumptions, e.g. that patterns in nature are necessarily hierarchical. But such a move would mean that transformed cladistics is unintelligible, and fails to present a coherent conceptual framework. This would also undermine the plea for methodological neutrality because some explanatory framework would be required for the purposes of justification. Alternatively, if the transformed cladist wants to avoid the charges of either Platonism or unintelligibility, then it can *only* be argued that his methods are rendered intelligible by evolutionary theory.

Models, methods and evolution

In examining the question of whether the model necessarily appeals to evolutionary assumptions, a basic point must be borne in mind. It has been argued, in the context of internal consistency in transformed cladistics, that it is not necessarily possible to reject outright the claim that homologies do not define hierarchies of groups, since this would presuppose that all hierarchical patterns found in nature are artefacts.[2] It is suggested that no cladist would argue against hierarchies existing in nature, because it is expected that independent workers do find them and corroborate them. But this argument fails to point out the reasons for the commitment to a hierarchical structure in the first place, or the sense in which hierarchical patterns in nature could be false.

Commitment to the hierarchy can be for several reasons. It can be because of commitment to evolutionary theory (as in evolutionary systematics and phylogenetic cladistics), or it can be due to a commitment to some other factor, such as information storage and retrieval (in phenetics). Pheneticists aim to produce repeatable classifications which satisfy the criterion of Gilmour naturalness. In Chapter 4, I argued that the strong pheneticists were dishonest to retain a hierarchical system purely for the purposes of speedy and convenient information retrieval. In contrast, the commitment to a hierachical system in transformed cladistics is not so clear out. They reject the phenetic rationale based on speed of information retrieval, and bypass the problem of information distortion. For example, phenetic measures of overall similarity, when expressed in a hierarchical classification, become distorted. For the transformed cladist there is no such problem because synapomorphies are specifically wedded to the hierarchy through internesting. They also reject the evolutionary basis for hierarchical classifications, arguing, instead, that they only search for hierarchical patterns in nature. What rationale is left for their committment to hierarchical patterns?

One possible reason is based on methodological considerations. While it may be empirically true that all we see in nature is an overwhelming diversity, and that, in fact, the representation of a hierarchy is an artefact of the method, e.g. in the sense that we do not directly 'see' the hierarchy, we would still have to explain how the methods of transformed cladistics can create the appearance of a hierarchy when there is none. Even if such an enterprise to discredit transformed cladistic methodology was successful, they could still argue that their model and methods were consistent. The search for hierarchical patterns would then be justified on methodological grounds. Unfortunately this strategy is undercut by the Platonic perspective implicit within transformed cladistics, since it could be argued that hierarchical patterns in nature are false, in the sense of representing 'appearances' of the underlying order. What, then, is the point of being committed to the hierarchy on methodological grounds, if there is the implicit assumption that such patterns are probably false? In this respect, the commitment to the hierarchy can only be arbitrary.

If the transformed cladists are to avoid this charge of arbitrariness, and given that we allow for the hierarchy, then the notion that such order is accidental rather than causally significant is without foundation, and it is at this point that appeal has to be made to evolutionary assumptions. In other words, both the model and methods are arbitrary *except* in the light of evolution. For example, the idea that patterns are discrete and hierarchical automatically introduces the assumption of homogeneity – that all groups included at a particular hierarchic level are similar in kind – implying that some causal factor is necessarily operating over these groups. This causal factor is most readily explicable in terms of evolutionary theory since similarities among hierarchically ordered organisms are the expected outcome of the evolutionary process itself. This does not mean that an evolutionary assumption is made here, only that the commitment to hierarchical patterns is hard to explain in transformed cladistics, except in terms of evolution. Similar problems are encountered in the transformed cladists' discussion of the concept of synapomorphy, where there is a tension between the characterisation of a timeless order and the methods employed to construct such an order.

In the last chapter (p. 152), we saw that to talk of synapomorphies in terms of more general and less general levels implicitly assumes *continuity* between characters. In regarding all characters as modifications of others, Platnick & Nelson remark on the possibility of a great chain of characters (or homologies) stretching from those of complete generality, which are true for all life, on to those true for a single species.[3]

Such a remark suggests a timeless order which is established on the basis of temporal considerations employed in the ontogenetic method. In one sense, it is obvious that the transformed cladists could be invoking Platonic assumptions, by appealing to an underlying order. Continuity between characters, in this respect, would imply contiguity: the very fact that characters can be linked in some way forms the basis of our perception of the one order. But, if this commitment to Platonism is rejected, then it is hard to envisage a timeless order which presupposes continuity between characters. Continuity implies 'movement' in a certain dimension, e.g. spatial, temporal, while a timeless order is, by definition, eternal and unchanging. No 'movement' occurs.

The difficulty in characterising a timeless order is increased by problems in specifying synapomorphies in non-evolutionary terms. Despite the claim that synapomorphies do not represent phylogenetic or functional factors, but are only defining characters of a group, there is a sense in which they can convey evolutionary implications. This concerns the method rather than the model. Hill & Crane characterise synapomorphies

as relatively idealised indications of organizational status rather than 'environmental adaptations',[4]

and while this reflects the point repeatedly made by transformed cladists that calling characters adaptations, complexes or apomorphies does not affect the method, this is not necessarily the same as calling such a method neutral with respect to evolutionary theory. There are surely instances where a defining character represents an environmental adaptation. To ignore such a possibility does not mean that such an assumption will disappear. Furthermore, transformed cladists distinguish between the phenotype and genotype, usually selecting their characters from the phenotype. Such a distinction is hard to explain except in terms of evolutionary theory.

Transformed cladists believe that if they ignore the processes that produced various patterns, then this automatically implies that their patterns are devoid of evolutionary content. The point at issue here is that while a strategy of methodological neutrality may remove all evolutionary considerations from the procedures of pattern analysis, it is hard to see how such methods can be justified except through recourse to an explanatory framework. In the absence of such a framework, the methods may well be arbitrary. This point can be more clearly seen in another fundamental aspect of cladistic analysis, the use of ontogenetic techniques for distinguishing between general and less general characters.

As mentioned in Chapter 5, ontogenetic techniques are direct techniques because they are based on observation alone. Transformed cladists further argue that ontogeny is both necessary and sufficient in the detection of hierarchical patterns, and has the advantage of eliminating additional hypotheses. For example, the use of outgroup analysis in the detection of hierarchical patterns requires a hypothesis of phylogeny at the next highest level, and therefore incorporates a time dimension. Furthermore, the use of ontogenetic information

> can be justified by reference to parsimony rather than recapitulation.[5]

For the present I will assume that the appeal to parsimony is justified.

In its original formulation, the ontogenetic method

> is based on Haeckel's law of recapitulation . . . the sequence of embryonic inductions in the ontogeny of an organ will usually parallel its phylogenetic development.[6]

Since this formulation incorporates evolutionary assumptions, transformed cladists prefer to use Von Baer's formulation since it is couched in terms of general and particular features:

> The further we go back in the development of Vertebrates, the more similar we find the embryos both in general and in their individual parts. . . Therefore, the special features build themselves up from a more general type.[7]

Nelson gives his own interpretation of this formulation:

> given an ontogenetic character transformation, from a character observed to be more general to a character observed to be less general, the more general character is primitive and the less general advanced.[8]

The hypothesis of the course of evolution, where, say, character *y* is more primitive than character *x*, is based on the observation that character *y* is more general than character *x*. Nelson's formulation does not necessarily presuppose any evolutionary assumptions:

> The concept of evolution is an extrapolation, or interpretation, of the orderliness of ontogeny.[9]

Ontogenetic techniques are also superior over other techniques because they are supposedly falsifiable. Techniques such as the paleontological method are not. However, Voorzanger & Van der Steen[10] have argued that Nelson's formulation of the biogenetic law in terms of generality is unfalsifiable because his terminology is confusing and requires background evidence (not relating to ontogeny). It is also worth noting that some characters do not have ontogenies, e.g. chromosome number, and that regeneration is a falsifier of Von Baer's law. However, such arguments are not crucial to my discussion of the possible arbitrariness of ontogenetic techniques that do not appeal to evolution, although a relaxation of Nelson's criterion of falsifiability would salvage the paleontological method. This, of course, presupposes that the reasons for rejecting paleontological evidence are dependent on falsification.

The transformed cladists are keen to emphasise that the ontogenetic method is empirical; it implies the development of similar individuals over generations and does not require any prior assumption of relationships:

> All one need assume is that since ontogenetic transformation is consist-
> ently observed to be in one direction, and never the reverse, we have
> direct evidence of transformations, and may rate the untransformed
> state as more general ('primitive') than the transformed state.[11]

It is clear, though, that this empiricism is based on the observation of differ-ent entities in time. In the assessment of character status there is always a necessary time dimension, whether it be in an ontogenetic or morphological transformation series. The search for ontogenetic novelties (recurrent patterns of characters) presupposes a relative time dimension because one character arises prior to another in the developmental sequence. This gives a hypothesis of *historical* ontogenetic transformations, since the pattern has a built in time component. This could imply that the resultant cladogram has a built in time axis. Yet, transformed cladists argue that the cladgram is an atemporal scheme, a timeless order.

However, this apparent contradiction is not necessarily fatal to the trans-formed cladists, because they would argue that only temporal considerations (i.e. methods which look at different entities in time) have been used in establishing an underlying timeless order. The temporal order that we see is but a manifestation of the underlying order. Under a Platonic interpretation this is plausible, but without it, it is hard to characterise a timeless order in empirical terms alone. On the one hand, transformed cladists imply that what is observable is, in a sense, unreal because a timeless order is abstracted from a method which incorporates temporal considerations. On the other hand, they also assert that what is observable, i.e. more general and less general characters, is real and definable. There is a tension here: how can the observables be real? Furthermore, if it is asserted that the assumption of con-tinuity through relative time is crucial to the ontogenetic method, then how can a timeless order be specified? Alternatively, if the aim is to find a time-less order, then what is the rationale for using ontogenetic methods to infer nesting of characters? Other methods, e.g. morphological series, could be equally appropriate in the search for levels of generality. If the transformed cladists argue that the idea of ontogeny recapitulating phylogeny is not problematic for their method, then neither is the abstraction of an order from morphological series using outgroup comparison problematic. Thus the methodological superiority of the ontogenetic method over the paleontological method cannot be argued on the grounds that the former incorporates a methodology free of all evolutionary assumptions. Both methods search for different entities in time, but this is not the same as one or other method incorporating evolutionary assumptions. If the ontogenetic method can specify a timeless order, then so can the paleontological method. In this respect, the choice of the ontogenetic method is arbitrary. Of course, the transformed cladists argue, in addition, that a fundamental defect of the paleontological method is its use of paraphyletic groups. But this is nothing more than an admission that for something to be real, it has to be definable, implying that the only reason for preferring ontogenetic methods is because they are useful in a Platonic framework.

In conclusion, if the transformed cladists assert that evolution is an extrapolation of the ontogenetic order, then they either resort to a Platonic interpretation, or characterise the timeless order using an arbitrary rationale for choice of method. If their claims are to make sense at all, they must have a basis in evolutionary theory. The choice of the ontogenetic method could then be justified in the light of ontogeny recapitulating phylogeny. Certainly this strategy would allay fears that ontogenetic novelties have no connection with evolutionary processes. For example, in arguing that the patterns obtained approximate to an idealised organisational status, transformed cladists fail to account for the fact that a developing animal has to function as an individual in the same way as an adult. Parallel adaptations to embryonic life are just as likely to be found here as they are in other life-stages. The use of the ontogenetic method will not screen against such possibilities.

To summarise, both the model and methods of transformed cladistics are only intelligible in terms of evolution. It is because the underlying assumption of homogeneity in hierarchical systems is predicated on continuity through time, that the subsequent distinction between less general and more general characters should be based on modification through time. This is most readily explicable in terms of evolution.

I now turn to the role of parsimony in transformed cladistics, since parsimony is invoked in many areas for justificatory purposes. Does an appeal to parsimony enable the transformed cladists to avoid either evolutionary or Platonic assumptions, while at the same time rendering their methods intelligible?

Parsimony and evolution

Parsimony is often regarded as an inevitable component of scientific method, reflecting an intuitive appeal to simplicity which underlies our approach to the formation of hypotheses. Both the phylogenetic cladists[12] and transformed cladists[13] justify the appeal to parsimony in this way. All cladists use parsimony in the construction of cladograms and not trees, emphasising the point that parsimony requires no assumptions about the contingent properties of the evolutionary process. But there are two separate applications of parsimony in systematics. In the first, parsimony is invoked in the reconstruction of phylogenetic patterns, where it is a method for inferring phylogenies. Parsimony finds that phylogeny in which the observed characters could have evolved with the least evolutionary change, thereby minimising requirements for *ad hoc* hypotheses of homoplasy.[14] In this sense, the parsimony principle is little more than a restatement of Hennig's[15] auxillary principle, in which homology should be presumed in the absence of evidence to the contrary. Secondly, parsimony is invoked in classification construction, where it can function as a measure of how simply information is represented in a classification. Thus, in phylogenetic and transformed cladistics it is the grounds for a justification of parsimony, as well as the assumptions made in the presence or absence of evolutionary considerations, that differ.

The idea that parsimony is simply a component of the scientific method cannot be established as easily as the cladists make out. In his book *Simplicity*, Sober[16] noted that parsimony is concerned with simplicity as a characteristic of hypotheses rather than a characteristic of phenomena, because the reasons for calling a linguistic description simple are different from the purposes for calling the phenomena it describes simple. In accordance with this, no phylogenetic cladist claims that evolution proceeds parsimoniously.[17] There are fundamental objections to the belief that nature is simple, not least because it is an insubstantiated empirical assumption. Molecular studies bear this out.[18] But even if our hypotheses do specify that nature is simple, it can still be argued that the sample of hypotheses that we have are not unbiased, for they represent just that aspect of nature which has been found sufficiently simple to be manageable to the human mind, and no conclusion can be drawn from them to the rest of nature.

Sober goes on to argue that simplicity is a secondary criterion by which we judge the relative merit of hypotheses, while other criteria such as logical consistency, part confirmation, absence of refutation and coherence with a wider domain of theory, taking precedence. Sometimes, though, it may be the case that only the addition of simplicity makes the set of criteria sufficient for unique choice. In these instances, simplicity of a scientific hypothesis is a matter of its informativeness, which in turn is a question of how much extra information has to be obtained to enable the hypothesis to answer questions of interest to the scientist. The less extra information required before the hypothesis answers the question, the more informative the hypothesis.[19]

An important point in Sober's account is that there exists a fundamental tension between simplicity, in terms of informativeness, and support (such that the information should be true). Intuitively simpler hypotheses are always objectively more informative in the sense that objective informativeness of a hypothesis is the measurable extent to which the hypothesis alone answers questions about entities in its domain. At the same time, such hypotheses must be well supported, implying that there is a trade off between simplicity and support. Although Sober gives no formal definition of support, Beatty & Fink[20] argue that parsimony considerations in systematics can be interpreted in terms of Sober's sense of informativeness:

> In systematics, questions of parsimony most commonly arise with regard to alternative cladistic (genealogical) accounts of the similarities and differences of several taxa. That is, data describing the similarities and differences of several taxa equally support a number of different hypotheses as to the genealogical relationships of the taxa.[21]

Given that we need some criterion other than support to decide among competing hypotheses, we should use parsimony.

> Generally, the hypothesis considered the most parsimonious is the one which, in conjunction with the fewest phenotypic evolutionary steps, accounts for the similarities and differences of the taxa in question.[22]

But this justification of parsimony is misguided; an analysis in terms of Sober's account (informativeness and simplicity) is inappropriate. As Sober himself states,

> The simplicity of a hypothesis is relative to a [body] of predicates; the support of a hypothesis is relative to a body of evidence.[23]

Because of this difference between simplicity and support, Friday[24] has argued that the appeal to parsimony in cladistic analysis is generally in connection with support rather than simplicity. This is certainly the sense that Farris[25] and Sober[26] imply. But,

> [i]f parsimony is part of the logic of support, does this mean that we can equate most parsimonious in the sense of least incongruence or fewest events, with most supported, and make an assessment of the goodness of our hypothesis?[27]

The response to this is a definite no, because an appeal to parsimony in connection with support presupposes that the simplicity of a hypothesis is evidence of its truth. In this respect, likelihood is standardly taken as a measure of evidential support. The likelihood ratio of two hypotheses relative to an observation indicates which of the two hypotheses is better supported by observation.[28] Phylogenetic cladists assume that parsimony and likelihood coincide: parsimony correctly records which cladistic groupings are best supported by observation. But such a conclusion is not justified because comparisons of minimum evolution and maximum likelihood methods do not necessarily give the same genealogical patterns using the same data.[29] The minimum evolution solution has a lower likelihood in cases where patterns differ. Thus, to defend the use of parsimony in phylogenetic inference on the grounds that it correctly identifies which genealogical hypothesis is best supported by observed character distributions is misguided.

What, then, can be said of the implicit assumptions that lie behind the use of parsimony in phylogenetic cladistic methods? Farris[30] has claimed that parsimony rests on realistic assumptions, not because it makes just the right supposition on the course of evolution, but because it avoids uncorroborated suppositions whenever possible. Parsimony presumes that evolution is not irreversible, that rates of evolution are not constant, and that all characters do not evolve according to identical stochastic processes. But in this account, there is one basic assumption which Farris ignores, that of incongruence. All phylogenetic cladists incorporate a model that is based on divergence, and use the principle of parsimony to specify the most congruent arrangement. Yet, whenever incongruency occurs, this in itself questions the hypothesis of perfect congruence. If perfectly congruent patterns cannot be found, then it is dishonest to take the best supported solution as the most preferred. Furthermore, where parsimony admits to error, this error is confounded with divergent change.[31] Some measure of congruence should be built into the model, otherwise the principle of parsimony will not only specify a method, but will also be part of the model. Thus, over and above the failure of parsimony methods to specify the best supported solution, lack of a model for evaluating distributions of non-divergent evolutionary events will impair any attempt to choose between competing solutions. In using a model and method which are interdependent, the phylogenetic cladists tend to be conservative, because they retain their hypotheses in the face of what can actually be observed, rather than using the principle of parsimony to search for convergence in conjunction with a model that has built in congruence.

From an evolutionary perspective, it is clear that the principle of parsimony counsels the rejection of parallelism and convergence so as to minimise the number of *ad hoc* hypotheses.[32] But this is somewhat arbitrary given the phylogenetic cladist's strict adherence to a model of divergence, and use of a method which searches for perfect congruence. Indeed, there are no real methodological constraints on such a procedure because we can always search for that cladogram which unites the largest number of apparent synapomorphies. 'True' synapomorphies are those that are consistent with all the others; but clearly, such synapomorphies may be false, in the sense of actually specifying a real case of convergence. The method only reaffirms the model.

These problems concerning simplicity and support are overlooked by many cladists (both phylogenetic and transformed) because they usually justify an appeal to parsimony through Popper's falsification.[33] Given Popper's views on the scientific method, they argue that general considerations about falsifiability demonstrate that parsimony is the correct method to use in systematics. Wiley characterises the most parsimonious genealogies as those which are least falsified on available evidence,[34] and Nelson & Platnick concur:

> an argument is parsimonious to the extent that it does not incorporate ad hoc items as protection against falsification.[35]

It should be obvious that the cladists are misguided in believing that general Popperian principles concerning the scientific method are sufficient to justify parsimony. In the first instance there is a disanalogy between Popperian methods and those of the cladists. In advocating that 'simple laws' should not be protected from falsification by *ad hoc* hypothesis, the cladists are misrepresenting simplicity in Popper. Simplicity in Popper[36] refers to the expression of hypotheses in a form where an attempt at falsification can be easily made; it does not refer to the construction of hypotheses chosen through parsimony. For Popper hypotheses are simple so that they can be falsified. This is not the same as falsifying all hypotheses and choosing the least falsifiable ones.

A further problem with Popper's account, and one which is not clarified by the cladists, is his tying together of support and simplicity. For Popper, the simplest hypothesis is the most falsifiable hypothesis. Simple hypotheses are more falsifiable in the sense that they permit (i.e. are compatible with) a smaller class of observation reports, and to the extent that simpler hypotheses are more falsifiable, simplicity is a characteristic of scientific aims. In this respect, Popper argues that support is a goal of the scientific method, since it is compatible with the goals of simplicity, informativeness and falsifiability. But as Sober has noted, content does not always correlate with our intuitions of simplicity and is not always the best measure of acceptability. Popper is mistaken in equating simplicity with falsifiability because he assumes that support and simplicity are one and the same. In contrast, for Sober, a simplicity comparison of two hypotheses is relativised to a question and its answer set, assumptions about comparative content and the existence of a set of accepted natural predicates as well as the logic of support. Given this deficiency in Popper's account, it is not surprising that many cladists have confused simplicity and support. It is also ironic that cladists appeal to Popper on behalf of rejecting *ad hoc* hypotheses, while at the same time failing to give any concrete measures of support to their hypotheses. Their appeal to parsimony is nothing more than a *post hoc* justification for their methods. Along similar lines, Panchen has argued that

> [b]efore [the] methodology [of phylogenetic cladists] can become trust-worthy . . . it will have to abandon its pretence to the hypothetico-deductive method and to parsimony as anything other than *ad hoc* methods of comparing alternative groupings.[37]

Enough has now been said concerning the justification of parsimony in phylogenetic cladistics to warrant a comparison with the issues involved in transformed cladistics. Transformed cladists, like the phylogenetic cladists, appeal to parsimony in terms of support and simplicity. In the former sense, parsimony is invoked as a test of homology, while in the latter it is a methodological principle. Both aspects will be discussed.

Patterson[38] defines homology as one of a number of characters used in defining natural groups. For him, the principle test of homology is congruence of character distributions. Furthermore, 'if several homologies characterize the same group, as hair, mammary glands, ear ossicles, left systemic arch, etc. characterize Mammalia, we have overwhelming evidence of order.'[39] As in phylogenetic cladistics, there is an underlying assumption – the ideal of perfect congruence. But while the phylogenetic cladist incorporates a model of divergence, the transformed cladist attempts to characterise a model which assumes character congruency to be an indication of order. This model is hierarchical, and the method demands that each defining character appears only once. However, the fact that incongruencies are found calls into question the ideal of perfect congruence. Transformed cladists ignore this problem, arguing instead, that incongruency can, in principle, be completely eliminated. This underlies their commitment to a true order. In effect, they assume that the simplicity of a hypothesis is evidence of its truth. But simplicity is not a measure of truth: if we take two hypotheses that seem to be equally supported by the evidence, then what is the reason to suppose that the simpler of the two is more likely to be true? If both hypotheses are equally supported in that the evidence favours both equally, then it is tautologous that each is likely to be true, and there would be no point in looking at support. Now, one way in which the transformed cladists can avoid this problem is to invoke a Platonic framework: the appeal to parsimony is justified by specifying the one true order. Character congruency is an indication of the underlying order, and any discrepancy between simplicity and support can be explained away by appealing to an underlying order which is poorly mirrored in our observed incongruencies. If this appeal to Platonism is discounted, then the transformed cladist must confront the problem of incongruency, because the hypothesis of perfect congruency is untenable in the light of perceived incongruities. One possibility is to argue for a hypothesis of the form 'character congruency holds except in the following *n* cases. . . ' But this is *ad hoc*. Generally, it would appear difficult to remove character incompatibility using parsimony, and then claim that an approximation to the true order has been found, without either questioning the status of the one order or rejecting the ideal of perfect congruency. If transformed cladists are to justify the use of parsimony it cannot be through an appeal to support.

The transformed cladists, like the phylogenetic cladists, give an ill-founded methodological formulation of parsimony which is justified in terms of Popper's criterion of falsifiability. They also appeal to the role of parsimony as a criterion; it is a standard by which something can be judged or decided. Parsimony has nothing to do with evolution: it is a standard of intelligibility.[40] Without the parsimony criterion we would never be able to make sense of nature, since there would be no reason other than personal whim or idiosyncracy for preferring one of a myriad of possible hypotheses. Indeed, parsimonious hypotheses can only be defended by the investigator without resorting to authoritarianism or apriorism.

For these reasons the transformed cladists feel justified in using parsimony as a methodological principle, which can be applied in several ways. For example, parsimony can be invoked in the choice of cladograms, because the most parsimonious cladogram requires one to argue away or neglect fewer characters. This has important implications for the descriptive power of hierarchical systems, because those classifications that are descriptively most informative, are those that allow character states to be minimised as efficiently as possible. As Panchen[41] has noted, the temptation here is to produce the most parsimonious cladogram, 'parsimonious' in the sense of minimising ranks. If all terminal taxa can be neatly paired off, to the exclusion of monotypic taxa, then a perfect cladogram will be obtained.

The implicit claim of the transformed cladists is that parsimonious representation does not presuppose any causal factor:

> The causal factor . . . is the cause of the agreement among the
> diagrams, not that it is the cause either of parsimony or of the
> diagram.[42]

Parsimony is only a procedure. But there is a fallacious assumption in this
attitude if it is the case that transformed cladists still want to retain some
notion of truth, e.g. Nelson & Platnick argue that judgement of truth is a
matter independent of the assumption that any cladogram may be presumed
to be true. For while we may appeal to simply hypotheses, this will pre-
dispose us to look at a sample of hypotheses which represent those which are
most easily accessible through our method. On this basis no conclusion can
necessarily be drawn about these hypotheses, implying that no notion of
'complex' truth is incorporated. In this manner, the transformed cladists
indirectly search for a representation of nature which is simple, so prejudicing
the way we look at nature.

Furthermore, to argue that the justification for parsimony is based on
intelligibility appears somewhat vacuous. While simpler hypotheses may be
more intelligible, this does not necessarily give protection from apriorism
and authoritarianism. To advocate the use of simple hypotheses alone, as
transformed cladists do, could easily be taken as an authoritarian attitude.
Nor is it clear in what sense parsimony renders nature intelligible in the
absence of recourse to an explanatory framework. From a biological perspective, the search for simple hypotheses of order which have little grounding in empirical reality, does not necessarily tell us anything of interest.
Within an evolutionary framework, the use of parsimony in detecting convergence is intelligible. But to plead for intelligibility in the sense that the
transformed cladists do, is little more than *ad hoc* justification.

Unfortunate consequences

I now want to discuss some unfortunate consequences of the desire
for methodological neutrality, reinforcing the point that the aims of the
transformed cladists are untenable and unsuitable. My discussion will build
upon the arguments that I set out in Chapter 3 concerning the rejection of
the fossil record in classification construction. As we saw there, all cladists
rejected fossil evidence in cladistic analysis because fossils were a hindrance
in the specification of cladistic relationships, while in cladistic classifications,
fossils led to problems in ranking and the piling up of categories. For the
transformed cladist, the aversion to fossils is much greater than in
phylogenetic cladistics because of the former's self-confessed
methodological neutrality. Since all inferences concerning evolutionary
processes should be purged from the method, and given that knowledge of

the evolutionary process can only be derived from fossils alone, then fossils must be rejected wholesale. Fossils are also supposedly irrelevant because they are not susceptible to ontogenetic techniques. While all the arguments that I presented in Chapter 3 are relevant here, there remains a further class of objections raised by the transformed cladists, which exemplify the methodological turn from phylogenetic to transformed cladistics. These objections concern the definition and recognition of natural groups in nature.

As I argued in Chapter 5, transformed cladists use definability as a criterion of group reality and assume the existence of a hidden order. At the same time, the use of such a bizarre criterion of reality automatically rules out any appeal to fossil groups, because they are paraphyletic. Extinct groups, like therapsids (mammal-like reptiles) and rhipidistians, are unreal; neither group includes all the descendants of a common ancestor. Of course, it is these very groups that are of interest in evolutionary systematics because they represent adaptively unified complexes or grades.[43] The transformed cladists' reply is to the effect that these groups express something about the process of evolution (i.e. anagenesis) rather than the pattern of character distributions. Such groups are merely 'timeless abstractions'[44] which facilitate 'adaptive storytelling'.

Over and above the problems relating to definition and reality, there are other reasons for rejecting arguments against paraphyletic groups. Firstly, in the context of fossils, the status of any group is dependent upon whether it is extinct or extant. Paraphyletic groups typically represent extinct groups. This is intuitively unsatisfactory, because if we take a hypothetical situation in which the mammals had died out shortly after their appearance in the fossil record, then the therapsids would represent a real group. Of course, this did not happen, but it only emphasises the logical end point of transformed cladistic reasoning: real groups are those groups which have survived to Recent times, and they are only real because they have survived. It is plainly ridiculous to assert that of all the taxa that have ever existed, some are more real than others. All taxa that have ever existed are equally real – it is only our ability to discriminate between extinct and extant taxa which is problematic. Transformed cladists appear to believe that the differences between extinct and extant groups are based on differences between observation and inference, when in actuality, the differences are based on varying degrees of certainty. Any inability to characterize extinct groups is surely a fault of the method, and not the data base.

A second serious flaw in the characterisation of paraphyletic groups is that it is possible to define such groups using cladistic methods. For example, the rhipidistia can be defined on the basis of dermal bone patterning.[45] This would seem to indicate that the arguments for rejecting fossils are further removed from empirical and methodological issues than the transformed cladists would have us believe.

In effect, the transformed cladists' desire to reject the fossil record on methodological grounds has been hardened into an attempt to specify the real order in nature from a narrow perspective. Ontological claims (over the existence of groups) have replaced methodological claims. If such ontological claims are coherent and backed up with sound reasoning then there may be some justification for rejecting the fossil record. But they are not; not only are the transformed cladists far from explicit in what they mean by a real group, but there is also no reason to suppose that fossil groups are unreal. Rather, the aversion to fossils is founded on a prejudice: the discovery of the one order is a simpler, more basic matter than any analysis of fossils permits. Rejecting fossil groups means that ontologically economical hypotheses can be constructed – hypotheses which postulate the existence of fewer entities. But there is no reason to suppose that a 'simpler' order is more real. The transformed cladists put themselves in the illogical position of concentrating on the topology of the phylogenetic tree of living organisms, while at the same time making a plea for the investigation of the real order in nature. Such methods can only give an artificial classification. Obviously transformed cladistics is too rigid for practical use, without necessarily avoiding some of the pitfalls present in other methods. If fossils cannot be incorporated into one classification, irrespective of the methodological requirements, then the classification is of limited appeal. Conclusively, the transformed cladists are guilty of using methodological arguments to cover up dubious ontological claims.

The rejection of explanation: pattern and process

The plea for methodological neutrality in transformed cladistics fails on two, related counts. First, if any reference to a Platonic world view is to be avoided, such methods are only intelligible in the light of evolution. Second, these methods have unfortunate consequences which can only weaken the justification for methodological neutrality with respect to evolutionary theory in the first place. These conclusions imply that transformed cladistic methods are, at best, only compatible with the weak version of the descriptive attitude. To a large extent, then, we have already answered the following questions: (*a*) whether methodological neutrality justifies a stance of theoretical neutrality in the finished product *vis-à-vis* evolutionary theory; (*b*) whether all hidden causes are rejected, in the negative. In the context of these conclusions, I want to return to the stronger version implicit within transformed cladistics, which hints at the discovery of recurrent patterns that do not necessarily presuppose an evolutionary explanation. Specifically, I want to examine the fundamental distinction between pattern and process.

Transformed cladists argue that knowledge of evolutionary history is independent of the existence and discovery of natural hierarchic systems, so that the investigation of patterns in the distribution of characters (taxonomy) is both independent of, and a necessary prerequisite to, any investigation of the evolutionary process. But, it is not clear what 'independent' refers to: is it independence of methods or results? In what follows, I will argue that the transformed cladists *do* imply both senses: that patterns alone are necessary in the methods of classification construction, and that the results accommodate patterns alone. However, this distinction between pattern and process is not nearly so clear cut in either the methods or results, as is made out.

Nelson & Platnick argue that patterns (the hierarchical order of life) are historically and logically prior to processes (mechanisms that generate patterns):

> A process is that which is the cause of a pattern. . . Pattern analysis is, in its own right, both primary and independent of theories of process, and is a necessary prerequisite to any analysis.[46]

What kinds of priority are being claimed here? In the first place, there is a historical claim; that analysis of pattern occurred historically in time before any construction of causal theories. In support of this, Leith[47] has argued that Aristotle's scale of nature is so like our present-day classifications, that it shows evolutionary theory to be unnecessary in the everyday work of the taxonomist. Similarly, Platnick has argued that

> When Linnaeus and his contemporaries first made their observations about spiders, and drew the conclusion that the group Araneae exists, they had little, if anything, that by modern standards can be considered a causal theory explaining the existence of such groups.[48]

Patterson[49] has concluded the pattern analysis represents a return to pre-Darwinian attitudes. While this claim for historical priority is generally true, in the sense that standing patterns of biological order were constructed and then accorded an explanatory input,[50] it is important to put this claim in perspective. While it may be true that before Darwin process was irrelevant to pattern analysis, this view is, to some extent, undercut by the observation that scientific methodology has not remained static over the last three hundred years. Scientific standards have changed. Thus, one might argue that rather than regarding patterns as problems which are in need of explanatory input (yet independent of it), a more appropriate strategy would be to search for many possible patterns, where some explanation is needed to find the correct pattern. Turning now to the claim of logical priority, it is harder to grasp precisely what the transformed cladists have in mind. Certainly, the priority of pattern over process is not logical: even though a conclusion to an argument does follow logically from its premises, this does not mean that scientists first think of the premises and only then come up with the conclusion. Nor can the transformed cladists be advocating ontological priority – the existence of patterns prior to the processes that produced them. Rather, the point being made is that it is possible to classify plants and animals into groups and hierarchies based only on a study of structure. Organisms can be classified solely on the basis of what is seen – there is no need to know how the factors arose in the first place. This represents the methodological priority of pattern over process. Such a claim, however, appears to have epistemic import for the transformed cladists, because it is only possible to construct theories on the basis of patterns. Thus, any theory of evolutionary descent must suggest a process by which the pattern of nature has arisen, although the patterns really came first in the sense that the study of things causal preceded the study of causes of things.

But this is a spurious claim. The transformed cladists are wrong if they assert that this is the only way of constructing processes, because results from other domains of inquiry can yield theories of process. Of course, it could be argued that in such domains, analyses of processes are based on patterns, but the point behind the transformed cladists' assertion is that they mistake accidental dependence of processes on patterns (state of the science) for a principle of dependence. For example, very little is known about the process of convergence at present; when species split, it is not known whether the subsequent degree of divergence is a function of similarity in structure, e.g. similarity of genetic code, or a function of ecological factors. It is because little is known about such processes, that at present pattern analysis is usually performed first.

Furthermore, I do not think that the links between methods of pattern analysis and construction of theories of process are so tenuous. Nelson & Platnick phrase the problem as follows:

> It is not generally possible, for example, to resolve synapomorphies by study of the genetic processes that produce them. And it seems doubtful if the study of process and the study of pattern need ever merge more than they do now.[51]

This is surely too extreme a case. In ontogeny, while it is not possible to uncover the relationship between developmental transformations and genetic processes, there are still more 'superficial' processes which are intertwined with the patterns. For example, the loss of terminal parts can easily occur in a developmental program, implying that some understanding of ontogenetic processes is a prerequisite to any analysis of pattern that aims to uncover the one order by resolving incongruencies. There has to be some basis for judging whether the true pattern has been discovered; a judgement that appeals to process.

More generally, if the transformed cladists are to avoid a Platonic interpretation, the recognition of natural groups should not be devoid of inferences based on processes. Transformed cladists argue that criteria for the recognition and assignment of taxa to natural groups are independent of any theories of process that explain the existence of relationships of these groups.[52] But the reasons for recognising such groups are not without theoretical prejudgement. If natural groups exist, then one is unlikely to recognise them by randomly chosen criteria. Criteria for the recognition of natural groups will be compatible with processes that produced these groups even when these processes are unknown. The recognition of natural groups does not escape from current theory; rather, theory and perception are intertwined, even if the same natural groups are recognised. Natural groups are not inviolable ontological units because the idea of what is 'natural' has changed.[53]

I do not wish to add anything more to this question of methodological priority since the arguments made earlier in this chapter back up this point. Instead, I will examine the interpretation of results in cladistic methods, and characterise some failings of transformed cladistics in this respect.

Platnick[54] has remarked that in the search for an explanation of patterns, biologists will doubtless go for a process of evolution as an explanation of the natural hierarchic system. Because patterns correspond to the apparent orderliness of life, and are a reflection of the apparent hierarchy of similarity among organisms comprising the biota, there is no input of causality. Causality is invoked only in theories of process,[55] implying that causal explanations can be decoupled from the results of pattern analysis. Platnick's implicit point is since these patterns relate to a limited number of solutions (or explanations), of which descent is only one factor which may predetermine pattern, it is possible to distinguish two types of patterns.

(1) There are those patterns which have developed historically, implying that cladistic analysis is a general comparative method applicable to all studies of interrelationships which are the product of modification by descent.

(2) There are those patterns in *general*, which may represent patterns of any sort, and have no necessary connection with hidden causes, evolutionary or otherwise.

The patterns referred to in (1) are clearly indicative of the position of phylogenetic cladistics. All phylogenetic cladists adopt the basic premise that life has evolved. Gaffney claims that the only empirical assumptions needed to carry out cladistic analysis are that

> evolution has occurred and that new taxa may be characterised by new features.[56]

While this characterisation of cladistic analysis is deficient, the point behind it is that pattern is divorced from process, because comparisons of the features of organisms predicted from theories of process with those actually found in nature, should occur.[57] For the phylogenetic cladist, hierarchically ordered similarities among organisms are the expected outcome of the evolutionary process itself. In contrast, the transformed cladists only assume that nature is hierarchically ordered. They wish to uncover general patterns of recurrent characters (internesting sets of characters), implying that their results are without theoretical implications, and do not necessarily contain any explanatory import: ' . . . a cladogram, as a summary of pattern, is not necessarily evolutionary.'[58] The justification for this search for recurrent patterns is based on the asumption that they can be used as an independent test of evolutionary theory.

There are several problems with these views of the transformed cladists which do not support any conception of general patterns. The overriding problem is how a pattern can be specified in non-causal terms, and whether any general patterns, if found, tell us anything of interest.

While it is obvious that to talk of patterns in general ignores any account of evolutionary phenomena such as parallelism, convergence, reverse mutations, or of any related subjects such as stratigraphical description, the transformed cladist is confronted by the problem of choosing or specifying what a general pattern is in the face of no causal input. Of course, the transformed cladist does argue that there is the one order which is specified by congruency of evidence. Thus

> 'Evolution' in the sense of 'coming into being' by itself is no causal principle of non-random agreement, for 'evolution' or 'coming into being' implies nothing about non-random agreements or lack of them.[59]

But, in the absence of cause, there is no *a priori* existence to different groupings. One pattern is just as valid as any other, irrespective of whether a particular pattern is the most easily perceived. A readily accessible pattern may not correspond to the one true order. For example, in the grouping of the salmon (A), the lungfish (B) and the cow (C), there is no reason to suppose that grouping (AC) is any more real than other possible groupings, (AB) or (BC). There must be some causal principle to justify the choice of grouping. Ball sums up the situation succinctly when he states that

> Order is not just a matter of that which is most easily immediately, or
> easily perceived; there may be hidden orders of greater significance.[60]

The choice of one order or grouping over another must be for some reason,
whether it be on non-explanatory grounds, e.g. ecological reasons, or with
respect to the relevance of a question of explanation. Yet, the transformed
cladists do not appear to give any reason, other than those which are compatible with a Platonic framework. Furthermore, it is not feasible for a
hypothetical 'agent' such as the perfect taxonomist to eliminate
incongruencies on an empirical basis alone: the perfect taxonomist must go
beyond empirical hypotheses or order (inductively based) to encompass
higher-order hypotheses which 'transcend' experience and suggest explanations of the data. Otherwise real incongruencies cannot be eradicated from
the supposed order. Conclusively, the plea for realism by all transformed
cladists, as manifest by the one order, is undercut by the rejection of all
causal implications.

One possible way to salvage the transformed cladist position is to suggest
that it is only a method for the identification or recognition of groups. No
causal input is required for the identification of natural groups as they
appear in nature. But this leaves us with the suspicion that transformed
cladistics is nothing more than stamp-collecting, a decidedly unscientific
procedure. The transformed cladists must remove this suspicion by stipulating
their explanatory framework.

The second problem that transformed cladists have to confront in the
interpretation of general patterns concerns the use of hierarchy. In attempting to question patterns which are the product of descent, they are doubting
the evidential basis of evolutionary theory. Thomas Huxley realised that the
classification of organisms and the description of the common structure of
groups is not based upon the evolutionary hypothesis; rather

> the fact that organisms can be classified hierarchically and that the
> common structure of taxonomic groups of organisms can be described,
> constitutes evidence for the hypothesis.[61]

Transformed cladists, too, recognise 'the fact that animal and plant species
fall into groups and that these groups form a nested series, or hierarchy'[62]
and that this constitutes evidential priority for Darwin's theory of evolution.
Yet, if they are questioning the evidential basis for evolution, it is odd that
they retain the hierarchy. Without evolutionary implications, the transformed cladists have not clearly specified their commitment to the hierarchy, implying that the true order is not necessarily hierarchical. Indeed, if
they are searching for odd recurrent patterns, then they should question the
idea of hierarchical ordering. There is no reason why the constraints on

synapomorphies cannot be loosened to allow for non-hierarchical ordering. At least some estimate could then be made of the validity of hierarchical ordering in terms of data distortion. This would help to allay fears that hierarchical ordering is an artefact of the method, i.e. that hierarchical patterns are accidental. Thus the transformed cladist should pre-empt those who claim that nature contains very different groups, because the idea of a single, well specified hierarchy in nature is *not* well supported – hybridisation and an array of dismayingly ordered sets of parallelism and convergence can be substantiated. At present, the transformed cladists are in no position to specify any limits to the possibility of discovering radically different patterns. Indeed, because of the interdependence of pattern and process, it is illogical to dismiss the processes whilst failing to disprove that the hierarchical patterns in nature are the result of evolution.

The transformed cladists do have a rejoinder to this; namely, that a refusal to interpret their results according to current theory is not necessarily injurious to science. A reluctance to interpret patterns beyond description which traces its linkage to other patterns is not objectionable in itself, since those who believe that transformed cladistics is at odds with current theories must find a means to discredit the empirical claim that hierarchical patterns are found in nature, or explain how they can justify their own theories in the face of evolutionary evidence. But again this view is undercut by the interdependence of pattern and process, implying that there is a built-in interpretation in the patterns. If such an interpretation is not appealed to, then how are such patterns intelligble?

A more promising avenue which the transformed cladists adopt suggests that their patterns are an independent test of current theory. Thus, systematic work as a whole, to the extent that it reveals hierarchy in nature which is *not* a methodological artefact, can be viewed as a simple test of the hypothesis of descent with modification (i.e. processes are tested in explaining past and present configurations of life), since this is the only theoretically testable hypothesis yet proposed which predicts such order. But in following the descriptive attitude, there is no *a priori* reason why patterns should be an independent test of processes. All that the descriptive attitude asserts is that the construction of patterns is an alternative approach which has no necessary links with theorising. Furthermore, as the transformed cladists are fully aware, it is possible to introduce *ad hoc* hypotheses to bolster up the theory under test, so that it may be difficult to specify when a pattern is a test of process. In this respect, a more comprehensive approach would be to formulate an alternative theory, and then compare the competing theories using empirical data. Nor is it true that such tests are independent. Because the transformed cladists do not necessarily sever all ties to hidden causes, their tests are not independent. While this does not detract from some form of testing in general (since in science tests are rarely independent) this does mean that they fail to eliminate all bias from their methods and results. Finally, there is no reason to suppose that processes should be tested by pattern analysis alone. It is perfectly reasonable to use other methods of testing. In micropaleontology, which utilises good stratigraphical data, phylogenetic inferences can be tested using sea-floor dating techniques derived from geophysics and continental drift. There is no need for pattern analysis in the first place. But, because the transformed cladists want a natural order while avoiding the perils of dubious evolutionary assumptions, they also discard essential techniques for testing, such as the fossil record. Without fossils, how are transformed cladistic methods testable with respect to the correct branching sequence?

In conclusion, outside of a Platonic framework, the transformed cladists are confronted with several difficulties in the construction and specification of process-independent patterns. Not only are the methods of pattern analysis more dependent on processes than is generally supposed, but also the interpretation of results is more critical than is made out. It is not possible to divorce the results of pattern analysis from the processes that generated them. More important, though, is the fact that the transformed cladists are questioning the evidence for evolution. As it stands, they appear to assume that the evidence is rejected if the explanation is rejected. This is clearly misguided, and if the transformed cladists did find odd, recurrent patterns, then this would question the evolutionary evidence. But this seems unlikely given their obsession with the one, true pattern. If any significant inroads are to be made into questioning evolutionary theory, a more sensible strategy would be to search for all sorts of different patterns, and then compare them with possible explanations. Finally, as in transformed cladistics, when a methodology is divorced from any explanatory theory, the effect becomes purely 'operational' (in the non-philosophical sense) and seems to have dubious scientific merit. Little of interest can be gained from the results.

Conclusions

One important feature of the recent debate in taxonomy has been its involvement with more general issues in evolutionary biology. In this respect, transformed cladistics has pride of place. While any discussion of the disputes within taxonomy in the context of overall advances in evolutionary thinking during the last ten years is beyond the scope of this book, a few cursory remarks concerning the controversy over transformed cladistics are in order.

Platnick sums up the contribution of transformed cladistics in glowing terms:

> What Hennig may well have done in general (and perhaps may have set out to do) is to demonstrate the inadequacy of the syntheticist paradigm, by showing us that we are hardly likely to achieve an understanding of the evolutionary process until we have achieved an understanding of the patterns produced by that process, and that even today we have hardly begun to understand.[63]

The transformed cladists are under no illusion as to their contribution.[64] But this is misleading because the transformed cladists appeal either to a Platonic framework or are forced to account for their views in evolutionary terms. In this latter respect, their classifications will be susceptible to the results of research into evolutionary processes – especially those involving speciation. The implication is that if there is an inadequacy in the syntheticist paradigm, it will not necessarily be demonstrated using pattern analysis. It is much more likely that discussions centring around analysis of punctuated equilibria and the independent possibility of species selection will lead to changes in the syntheticist paradigm.

One interesting point that arises out of the discussion of the descriptive attitude is the difference between phenetics and transformed cladistics. Overall, strong phenetics is much more extreme than transformed cladistics. Yet, in the mid 1960s there was no real controversy over the status of Neo-Darwinism, as there is today. This would seem to suggest that the transformed cladists have, to an extent, cashed in on the current controversy that is present in other fields, e.g. structuralist approaches in developmental biology,[65] adaptationist critiques of natural selection, theories of macroevolution[66] and the recent theory of molecular drive.[67] In the absence of these changes, it is doubtful whether transformed cladistics would have caused so much controversy in taxonomy. Indeed, cladistics as a whole might have tended to concentrate more on empirical aspects,[68] than methodological procedures. Furthermore, the impropriety of using methodological rules that were anti-good biology might have been more evident. Less attention would also have been paid to philosophical justifications.

The unrest in recent debates over the status of Neo-Darwinism has tended to obscure a more important problem that is implicitly confronted in transformed cladistics. Do we want a taxonomic system in building a theory? If we do not, then from the standpoint of the Darwinian, the claims of the taxonomist should be ignored. If we do, then the transformed cladist must first substantiate why this is so, if he is to justify a questioning of Neo-Darwinism using taxonomic evidence.

In the final analysis, it is difficult to understand fully any viable rationale for transformed cladistics. Although transformed cladists may have set out to demonstrate the dependence of conventional taxonomy on Darwinian thought, and more radically, to unload evolutionary theory from taxonomic practice, their strategy is naive. If the true motivation for transformed cladistics is to search out replacement theories for Neo-Darwinism, then a competing theory should first be developed, and then tested in the face of empirical evidence. To first distance the empirical evidence from the underlying theory by methodological strictures, and then appeal for the rejection of theoretical input into taxonomic procedure on the grounds that it is an unscientific method, is poor logic and is certainly unnecessary in the face of the rapidly expanding fields of molecular genetics and developmental biology. In time these disciplines will provide data and hypotheses which will enable taxonomy to explore fully the interdependence of patterns and process.

Writing as a philosopher of science, perhaps the most distressing aspect of this controversy in taxonomy is the wholly misplaced appeal to the philosophy of science. While all cladists alike have been eulogising the values of Popperian falsification as an example of 'modern' thinking, nothing could be more wrong. Over the past twenty years the philosophy of science has experienced a rapid shift away from the type of hypothetic-deductivism expressed by Popper, and has had to 'take on board' a wide range of views such as methodological relativism, anti-realism, and the new pragmatism articulated by Rorty. Equally baffling is the transformed cladists' aversion to theories of process at a time when greater understanding of evolutionary processes is within reach. If anything, I think that transformed cladistics is a retrenchment against novel ideas that are emerging in evolutionary biology in general, and that the proponents are often more entrenched than those they wish to attack. If the transformed cladists want to practice what they preach, they will have to shed the veil of 'methodological neutrality' and pursue radically different patterns in nature as an end in itself, somewhat akin to the structuralist credo initiated by Lévi-Strauss.

NOTES

Introduction

1 For a discussion of such arguments, see 'How true is the theory of evolution', Editorial in *Nature*, **290**:75–6 (1981); 'Evolution: myth, metaphysics, or science?', J. Little (1980) in *New Scientist*, **87**:708–9; (1980) 'Evolutionary theory under fire', R. Lewin in *Science*, **210**:883–7.

2 The most comprehensive account of creationist thinking is given by Gish, D.T. (1978) *Evolution – The Fossils say No!*. San Diego: Creation Life Publishers. Michael Ruse (1982) gives a comprehensive reply in his book *Darwinism Defended. A Guide to the Evolution Controversies*. London: Addison-Warley; as does Kitcher, P. (1982) *Abusing Science. The Case against Creationism*. Cambridge, Mass.: MIT Press.

3 The most obvious example of such coverage can be seen in the debate over possible influences in exhibits at the British Museum (Natural History). Wade summed up the situation thus: 'Dinosaur battle erupts in British Museum. Anti-cladist sees reds under fossil beds in alliance with creationists to subvert the establishment', (in (1981) 'Dinosaur battle erupts in British Museum' – *Science*, **211**:35–6). An extensive correspondence on the influences of transformed cladists on museum policy can be found in *Nature*, **288, 289, 290** (1980, 1981).

4 Simpson, G.G. (1961) *Principles of Animal Taxonomy*. New York: Columbia Univ. Press, p.7.

5 *Ibid.*, p.11.

6 Hull, D.L. (1970) 'Contemporary systematic philosophies' in *Ann. Rev. Ecol. & Syst.*, 1:22.

7 Jardine, N. (1969) *Studies in the Theory of Classification*. Ph.D Thesis, Cambridge University, p.205.

8 The growth of cytogenetics and microevolutionary studies produced the area of study called experimental taxonomy.

9 Population genetics involves the study of the effects of natural and artificial selection, and related problems.

10 Sokal, R.R. & Sneath, P.H.A. (1963) *Principles of Numerical Taxonomy*. San Franciso: W.H. Freeman.

11 Simpson, G.G. *Principles of Animal Taxonomy.*
12 Mayr, E. (1969) *Principles of Systematic Zoology.* New York: McGraw-Hill.
13 Hennig, W. (1966) *Phylogenetic Systematics.* Urbana: Univ. Illinois Press.
14 Brundin, L. (1966) 'Transarctic relationships and their significance as evidenced by chironomid midges' in *Kungla. Svenska Ventensk. Akad. Handlingar*, 11:1–472.
15 Nelson, G.I. & Platnick, N.I. (1981) *Systematics and Biogeography.* New York: Columbia Univ. Press.
16 Patterson, C. (1980) 'Cladistics' in *The Biologist*, 27:234–40.
17 Platnick, N.I. (1979) 'Philosophy and the transformation of cladistics' in *Syst. Zool.*, 28:537–46.
18 It would not be unreasonable to recognise a 'movement' in taxonomy. The sense of 'movement' implied here is not sociological (a series of actions and endeavours of a body of persons for a special object) but structural (having a distinctive structure of its own). For instance, within taxonomy, there is a dynamic momentum, whose development is determined by its intrinsic principles as well as by the structure of the territory it encounters (e.g. the results of closely related fields of inquiry). Furthermore, like a stream it comprises several parallel currents, which are related but by not means homogeneous, and may move at different speeds. Finally, these currents have a common point of departure, but need not have a definite and predictable joint destination; it is compatible with the character of a movement that its components branch out in different directions.
19 For instance 'Evolutionary trends and the phylogeny of the agnatha' – L.B. Halstead (1982) *Problems of Phylogenetic Reconstruction.* London: Academic Press.

Chapter 1 *Theoretical and descriptive attitudes in taxonomy*

1 Gilmour, J.S.L. (1937) in *Nature*, 139:1040–2.
2 Sneath, P.H.A. & Sokal, R.R. (1973) *Numerical Taxonomy.* San Francisco: W.H. Freeman.
3 For example, see Gaffney, E.S. (1975) 'An introduction to the logic of phylogeny reconstruction' in *Phylogenetic Analysis and Paleontology*; Wiley, E.O. (1975) 'Karl R. Popper, systematics and classification – a reply to Walter Bock and other evolutionary taxonomists' in *Syst. Zool.*, 24:233–43.
4 Brundin, L. (1972) in *Zool. Scripta.*, 1:119.
5 Agassiz, L. (1859) *An Essay on Classification.* London: Longman, Brown, Green, Longmans, and Roberts and Trubner & Co., p.31.
6 Hull, D.L. (1970) in *Ann. Rev. Ecol. & Syst.*, 1:19–54.
7 Such an assumption is fairly prevalent in both phylogenetic cladistic and evolutionary systematic writings.
8 Hennig, W. *Phylogenetic Systematics.*
9 Eldredge, N. & Cracraft, J. (1980) *Phylogenetic Patterns and the Evolutionary Process.* New York: Columbia Univ. Press, p.17.
10 *Ibid.*, p.21.
11 Mayr, E. (1968) 'Theory of biological classification' in *Nature*, 220:545–8; also in *Principles of Systematic Zoology.*
12 Mayr, E. (1976) *Evolution and the Diversity of Life: Selected Essays.* Cambridge, Mass.: Belknap Press of Harvard, p.427.

13 Mayr, E. *Principles of Systematic Zoology*, p.79.

14 Simpson, G.G. *Principles of Animal Taxonomy*.

15 Mayr's Biological Species Concept is defined as 'groups of actually interbreeding populations which are reproductively isolated from other such groups,' in *Principles of Systematic Zoology*, p.26.

16 Simpson's Evolutionary Species Concept is defined as a 'lineage (an ancestral-descendant sequence of populations) evolving separately from others and with its own unitary role and tendencies,' in *Principles of Animal Taxonomy*, p.153.

17 Mayr also uses practical criteria for the recognition of higher taxa, such as gaps, and as we shall see later, this is not necessarily consistent with his general attitude that classifications are theories.

18 Ashlock, P.D. (1979) 'An evolutionary systematist's view on classification' in *Syst. Zool.*, **28**:446.

19 An alternative 'authoritarian' approach is exemplified in Troll's recognition of homology. (See Troll, W. (1926) *Goethes Morphologische Schriften*. Jena: E. Diederichs Verlag.) This is done by means of the conception of 'types', the real nature, the essential similarities and the real relationships. Both the discovery and verification of types is performed by intuition, in a coincidence (in the physical sense) of the subject (the investigator) and the object (that which is to be investigated). In this esoteric act of coincidence (*Schau*) types are seen (*geschaut*). Thus, the real nature is recognised by personal conviction and the feeling of having insight into the real order of nature. Such blatant authoritarianism was a major failing of the typological approach, although most scientific practices are never entirely free of this problem.

20 Typical definitions are phrased accordingly: e.g. De Beer's 'The sole condition which organs must fulfill to be homologous is to be descendants from one and the same representative in a common ancestor.' (Quoted from De Beer, G.R. (1928) *Vertebrate Zoology*. London: Sidgwick & Jackson, p.478.) Or, in Haas, O. & Simpson, G.G. (1946) 'Analysis of some phylogenetic terms with attempts at redefinition' in *Proc. Am. Phil. Soc.*, **90**:325: 'Homology, as we agree, is best defined as similarity interpreted due to common ancestry.'

21 Hull, D.L. (1967) 'Certainty and circularity in evolutionary taxonomy' in *Evolution*, **21**:174–89.

22 Sokal, R.R. & Sneath, P.H.A. *Principles of Numerical Taxonomy*.

23 I will take typological thinking to be the assumption that every natural group of organisms, and hence every taxon in a classification, has an invariant generalised or idealised pattern shared by all the members of the group. This idealised pattern is known as the *Bauplan*. Representatives of the old typological school who are frequently cited include Louis Agassiz (1807–1873) and Richard Owen (1804–1892); both emphasised the abstraction of an ideal pattern from the diversity of living forms. (After Ospovat, D. (1981) *The Development of Darwin's Theory*. Cambridge Univ. Press.)

24 Charig, A.J. in *Problems of Phylogenetic Reconstruction*.

25 A variation on this theme is presented in Hull's essentialist and non-essentialist dichotomy. Here, natural groups are said to exhibit 'essences'. See Hull, D.L. (1965) 'The effect of essentialism on taxonomy – two thousand years of stasis' in *Brit. J. Phil. Sci.*, **15**:314–26; **16**:1–18. Cf. Hull's account with Mayr's

definition of typological, 'Typological versus population thinking' in *Conceptual
Issues in Evolutionary Biology*, ed. E. Sober (1984) Cambridge, Mass.: MIT Press.

26 Hennig, W. *Phylogenetic Systematics*.

27 Simpson, G.G. *Principles of Animal Taxonomy*.

28 Kiriakoff made the distinction between typological and phylogenetic systematics
(In Kiriakoff, S.G. (1959) 'Phylogenetic systematics versus typology' in *Syst.
Zool.*, 8:117–8). In the typological approach three assumptions were implicit in
the method: (1) measures of similarity between taxonomic groups are a direct
indication of recency of common ancestry; (2) a phylogeny can be constructed
without fossil evidence; (3) classifications should be based on primitive, stable,
conservative, non-adaptive characters, rather than plastic, adaptive characters.

For Kiriakoff, these assumptions were seen to be representative of the old
typological school of Agassiz and Owen, and both evolutionary systematics and
phenetics were taken to be the modern day equivalents. In contrast, the
phylogenetic approach rejected outright the assumptions of typological systematics, and argued for (1) the lack of correspondence between similarity and
blood relationship; (2) the importance of paleontological evidence in
phylogenetic reconstruction; (3) the selection of unique derived characters. In
this scheme, only classical Hennigian cladistics fulfilled all the criteria necessary
for a phylogenetic approach. For the other side of this debate see Bigelow,
R.S. (1956)'Monophyletic classification and evolution' in *Syst. Zool.*, 5:145–6,
and (1958) 'Classification and phylogeny' in *Syst. Zool.*, 7:49–59.

29 Nelson, G.J. (1970) 'Outline of a theory of comparative biology' in *Syst. Zool.*,
19:374.

30 Nelson quotes Simpson, G.G. *Principles of Animal Taxonomy*, p.50.

31 This point is certainly implicit in Kiriakoff's shrewd observation that 'most
biologists who consider themselves to be phylogeneticists are typologists' in
Syst. Zool., 8:118 (1959).

32 Thus, the idea of the world and all its included objects must have existed
before the world itself.

33 Owen, R. (1849) *On the Nature of Limbs*. London: John Van Voorst, pp.84–6.

34 Agassiz, L. (1857) 'Essay on classification' in *Contributions to the Natural History
of the United States*, Vol. 1. Boston: Little, Brown & Co., p.136; reprinted 1962,
ed. E. Lurie. Cambridge, Mass.: Harvard Univ. Press.

35 Owen, R. *On the Nature of Limbs*, p.86.

36 Hesse, M.B. (1974) *The Structure of Scientific Inference*. The Macmillan Press.

37 *Ibid.*, p.45.

38 *Ibid.*

39 Wittgenstein, L. (1953) *Philosophical Investigations*. Trans. G.E.M. Anscombe.
Oxford: Blackwell, sec. 66, 67.

40 Hesse, M.B. *op. cit.*, p.46.

41 Beckner, M. (1959) *The Biological Way of Thought*. New York: Columbia Univ.
Press. All quotes from 1968 edition (Univ. Calif. Press, Berkeley).

42 Because the terms 'monotypic' and 'polytypic' have had varied meanings, I will
adopt the corresponding terms 'monothetic' and 'polythetic' to avoid confusion.

43 Beckner, M. *The Biological Way of Thought*, p.22.

44 Jardine, N. (1969) 'A logical basis for biological classification' in *Syst. Zool.*, **18**:37-52.

45 See Farris, J.S. (1982) 'Simplicity and informativeness in systematics and phylogeny' in *Syst. Zool.*, **31**:413-44.

46 For a general discussion see Hesse, M.B. (1974) *The Structure of Scientific Inference*, pp.9-49; Sellars, W. *Science, Perception and Reality*. London: Routledge & Kegan Paul.

47 Initially this distinction was made at the level of predicates, but was later modified to the level of sentences.

48 Putnam, H. (1962) 'What theories are not' in *Logic, Methodology and Philosophy*, ed. E. Nagel. Stanford, California: Stanford Univ. Press.

49 *Ibid.*, p.240.

50 Suppe, F. (1973) 'Facts and empirical truths' in *Can. J. Phil.*, **3**:197-212.

51 Beatty, J. (1983) 'What's in a word? Coming to terms in the Darwinian revolution' in *Nature Animated*, ed. M. Ruse. Dordrecht: D. Reidel, pp.79-100.

52 Churchland, P.M. (1975) 'Two grades of evidential bias' in *Phil. Sci.*, **42**:250-9.

53 *Ibid.*, p.251.

54 *Ibid.*

55 Goodman, N. (1954) *Fact, Fiction and Forecast*. London: Athlone Press.

56 Sattler, R. (1966) 'Towards a more adequate approach to comparative morphology' in *Phytomorphology*, **16**:417-29. See also his (1967) 'Petal inception and the problem of pattern detection' in *J. Theor. Biol.*, **17**:31-9.

57 These terms refer to various categories of plant-shoot construction.

58 See p.27 for definitions of these terms.

59 For instance, in the transformation of cartilage to bone, is the change continuous or discrete?

60 Zangerl, R. (1948) 'The methods of comparative anatomy and its contribution to the study of evolution' in *Evolution*, 2:351-74.

61 Zangerl also distinguishes between experimental and non-experimental methods. The results of the former cannot be used to reject findings from the latter.

62 After Zangerl, R. *op. cit.*, p.352.

63 *Ibid.*, p.355.

64 *Ibid.*, p.357.

65 Although Zangerl does admit that morphology, like all other sciences, has its own domain of theoretical evaluation of results as manifested in comparing morphological theories with phylogenetic theories. However, the two types of theory are not the same because they do not rest on the same *erkenntnistheorisch* plane (i.e. level of cognition).

66 Boyden, A. (1973) *Perspectives in Zoology*. Oxford: Pergamon Press, p.82.

67 Bock, W.J. (1974) 'Philosophical foundations of classical evolutionary classifications' in *Syst. Zool.*, 22:386-7.

68 Hennig, W. *Phylogenetic Systematics*, p.93.

69 'Homology is the study of monophyly of structures' – suggested by Hecht, M.K. & Edwards, J.L. (1977) 'The methodology of phylogenetic inference above the species level' in *Major Patterns in Vertebrate Evolution*, eds. M.K. Hecht, P.C. Goody & B.M. Hecht. New York: Plenum Press.

70 Eldredge, N. & Cracraft, J. *Phylogenetic Patterns and the Evolutionary Process*, p.36.
71 Jardine, N. (1967) 'The concept of homology in biology' in *Brit. J. Phil. Sci.*, **18**:130.
72 Sokal, R.R. & Sneath, P.H.A. *Principles of Numerical Taxonomy*, p.71.
73 Hull, D.L. (1968) 'The operational imperative: sense and nonsense in operationalism' in *Syst. Zool.*, **16**:438–57.
74 Sneath, P.H.A. & Sokal, R.R. *Numerical Taxonomy*, p.77.
75 Patterson, C. in *Problems of Phylogenetic Reconstruction*, p.65.
76 Patterson hints at these attitudes in his account of homology. He distinguishes between: 'The taxic approach . . . is concerned with monophyly of groups. The transformational approach is concerned with change, which need not imply grouping,' (*Ibid.*, p.34). The taxic approach leads to a hierarchy of groups, while the transformational approach does not.
77 Structural fit need not be entirely based on similarity. One could alternatively have correspondence in functional development.
78 Van Valen, L.M. (1982) 'Homology and causes' in *J. Morphol.*, **173**:305–12.
79 Hence Van Valen's definition: 'Homology is resemblance caused by a continuity of information,' (*Ibid.*, p.305).
80 Jardine, N. (1969) 'The observational and theoretical components of homology: a study based on the dermal skull-roofs of rhipidistian fishes' in *Biol. J. Linn. Soc.*, **1**:327–61.
81 Jardine, N. & Sibson, R. (1971) *Mathematical Taxonomy*, London: John Wiley, p.320 (my italics).
82 Jardine, N. (1969) in *Biol. J. Linn. Soc.*, **1**:335–7.
83 Jarvik, E. (1948) in *Kungla. Svenska. Vetensk. Akad. Handlingar.*, **25**:1–301.
84 An assumption which is clearly false.
85 Jardine, N. (1969) in *Biol. J. Linn. Soc.*, **1**:357.
86 Patterson determines 'special' similarity on the basis of correspondence in relative position, with additional criteria derived from similarity in developmental studies. However, he also regards congruency as a principle test of homology.
87 An interesting analogy can be drawn with astronomical theory, in which practical considerations are kept much further apart from theory, than in homology. In astronomy there is a separate theory of instruments, based on laws of properties of lenses and on the refractive indices of such lenses. The importance of developing a separate theory of telescopic vision apart from an understanding of celestial phenomena was recognised by Galileo himself (see Feyerabend, P.K. (1975) *Against Method*. London: Verso). In the following scheme

                                                    theory

    theory of   (1)
    instruments   (2)   observation
    _____

theory and observation (1) are distinct from theory of instrumentation (2). In homology, the ties between the practical component and theoretical component are much closer.
88 'When two structures are described as homologous (or homologues), an abso-

lute statement is made about their identity, after which different states of such homologues may be used as wholly equivalent in constructing a classification,' in Inglis, W. (1966) 'The observational basis of homology' in *Syst. Zool.*, **15**:219.

89 In genetic data this presents no problem, since only discrete entities are used. But there are other difficulties. If we wish to homologise two binary sequences, at what stage can we say that they are homologous? This is not easy, because genome DNA includes large areas of repetitive DNA (untranslated, non-coding DNA) as well as translated DNA (structural genes). How much importance should we attach to either? Should we go for overall matching or just matching within a particular sequence (e.g. structural DNA)? Both options presuppose a decision for comparison, and may give very different results.

90 Sattler, R. (1966) in *Phytomorphology*, **16**:417–29.

91 Thompson, D'Arcy W. (1942) *On Growth and Form*. Cambridge Univ. Press, 2nd ed., p.14.

92 *Ibid.*, pp.1078–9.

93 After Sneath, P.H.A. (1967) 'Trend-surface analysis of transformation grids' in *J. Zool.*, **151**:67.

94 Woodger explains this succinctly: 'When we compare two things we set up a one-to-one relation or *correspondence* between the parts of one and those of the other and proceed to state how corresponding parts resemble or differ from one another with respect to certain sets of properties.' Quoted from Woodger, J.H. (1945) 'On biological transformations' in *Essays on Growth and Form*, eds. W.E. Le Gros Clark & P.B. Medawar. Oxford: Clarendon Press, p.98.

95 Thompson, D'Arcy W. *On Growth and Form*, p.1032.

96 Woodger, J.H. *op.cit.*, p.109.

97 *Ibid.*

98 For example, in the aforementioned conversion, there is a lack of distinction in the shift from a morphogenetic pattern to a phylogenetic pattern. The assumption of continuity of Bauplan being equivalent to community of descent, accompanied by gradual modification of structure which leaves the Bauplan still discernible, is not always true. In some cases it could be argued that the application of the evolutionary postulate is inapplicable.

99 Gould, S.J. & Lewontin, R. (1979) 'The spandrels of San Marco and the Panglossian paradigm: a critique of the adaptationist programme' in *Proc. Roy. Soc. Series B*, **205**:581–98.

100 Of course, a pre-Darwinian explanation could be given, and it is ironical that evolutionists, worried by this possibility, have used d'Arcy Thompson's example of directed change of proportions along single gradients, as the most pronounced case of directed evolution known.

101 Sneath, P.H.A. (1967) in *J. Zool.*, **151**:65–122.

102 Sokal, R.R. & Sneath, P.H.A. *Principles of Numerical Taxonomy*, p.81.

103 Thom, R. (1975) *Structural Stability and Morphogenesis: an outline of a general theory of models*. Trans. D.H. Fowler. Reading, Mass.: W.A. Benjamin.

104 As opposed to *global* descriptions, which would be description unconfined to a particular time or place.

105 The strong form of the descriptive attitude in taxonomy also utilises a similar notion of *local* description.

106 The study of geometrical properties and spatial relations unaffected by continuous change of shape or size of figures.

107 Thom, R. *Structural Stability and Morphogenesis*, pp.6–7.

108 *Ibid.*, p.8 (Thom's italics).

109 *Ibid.*, p.151 (Thom's italics).

110 *Ibid.*, p.159.

Chapter 2 *Evolutionary systematics and theoretical information*

1 Simpson, G.G. (1976) 'The compleat paleontologist' in *Ann. Rev. Earth Planet Sci.*, **4**:1–13.

2 Bock, W.J. (1977) 'Foundation and methods of Evolutionary Classification' in *Major Patterns in Vertebrate Evolution*, p.853.

3 Simpson, G.G. *Principles of Animal Taxonomy*, p.110.

4 My account here is taken from Simpson, G.G. *Principles of Animal Taxonomy*.

5 Evolutionary change along a branch.

6 I take this expression from Jardine, N. & Sibson, R. *Mathematical Taxonomy*.

7 See Mayr's *Principles of Systematic Zoology* and (1974) 'Cladistic analysis or cladistic classification?' in *Z. Zool. Syst. Evol.-forsch.*, **12**:94–128.

8 Cain, A.J. (1959) 'Deductive and inductive processes in post-Linnaean taxonomy' in *Proc. Linn. Soc. Lond.*, **170**:185–217.

9 Cain, A.J. & Harrison, G.A. (1960) 'Phyletic Weighting' in *Proc. Zool. Soc. Lond.*, **135**:1–31.

10 Mayr, E. *Principles of Systematic Zoology*, p.218.

11 Jardine, N. & Sibson, R. *Mathematical Taxonomy*.

12 Evolutionary systematists present three pieces of information in their phylogenetic diagrams or trees: degree of difference (abscissa), geologic time (ordinate) and degree of divergence (angle of divergence at branch points on the diagram). It must be noted, however, that there is no mathematical relationship between the degree of divergence in an evolving taxon and the angle of divergence as exemplified in the diagram. The relationship is purely arbitrary.

13 After Mayr, E. (1974) in *Z. Zool. Syst. Evol.-forsch.*, **12**:444, Fig. 1. (Reprinted in *Evolution and Diversity of Life: Selected Essays*. Cambridge, Mass: Harvard Univ. Press (1976), q.v. pp.433–76.)

14 After Dobzhansky, Th., Ayala, F.J. Stebbins, G.L. & Valentine, J.W. (1977) *Evolution*. San Francisco: W.H. Freeman, p.236.

15 After Huxley, J.S. (1959) 'Clades and Grades' in *Function and Taxonomic Importance*, ed. A.J. Cain. London: Syst. Assoc. Publ., Vol. 3. See also Huxley's (1959) discussion in 'Evolutionary processes and taxonomy with special reference to grades' in *Uppsala Univ. Arssk.*, **6**:21–39.

16 Simpson, G.G. *Principles of Animal Taxonomy*, p.124.

17 Of course, the other end of the spectrum would be pointless.

18 Simpson, G.G. (1951) 'The species concept' in *Evolution*, 5:286.

19 Simpson, G.G. *Principles of Animal Taxonomy*, p.189.

20 *Ibid.*, p.215, Fig. 29.

21 Simpson, G.G. (1963) 'The meaning of taxonomic statements' in *Classification and Human Evolution*, ed. S.L. Washburn. Chicago: Aldine, pp.1–31.

22 *Ibid.*, p.21.

23 Mayr, E. *Principles of Systematic Zoology*, p.70.

24 Mayr, E. *Evolution and the Diversity of Life*, pp.425, 427, 439. In a later account ('Biological classification: toward a synthesis of opposing methodologies' in *Science*, **214**:510–16 (1981)) this view has been watered down to 'Biological classifications . . . serve as the basis of biological generalisations.'

25 Mayr, E. (1968) in *Nature*, **220**:545–8.

26 Bock echoes Mayr in arguing 'that the most important purpose of biological classification is to serve as a hypothesis generating mechanism with stress placed on deductive hypotheses rather than inductive ones.' (See Bock, W.J. in *Major Patterns in Vertebrate Evolution*, p.865).

27 Advocates of this attitude include Eldredge, N. & Cracraft, J. *Phylogenetic Patterns and the Evolutionary Process*; Farris, J.S. (1982) in *Syst. Zool.*, **31**:413–44; Forey, P. 'Neontological versus paleontological stories' in *Problems of Phylogenetic Reconstruction*.

28 Brundin, L. (1972) in *Zool. Scripta.*, 1:114.

29 Mayr, E. (1974) in *Z. Zool. Syst. Evol.-forsch.*, 12:95.

30 Mayr, E. (1965) 'Numerical phenetics and taxonomic theory' in *Syst. Zool.*, **14**:79.

31 Patterson, C. (1981) 'Significance of fossils in determining evolutionary relationships' in *Ann. Rev. Ecol. & Syst.*, 12:195–223.

32 Wiley, E.O. (1981) *The Theory and Practice of Phylogenetic Systematics*. New York: John Wiley.

33 Mayr, E. (1974) in *Z. Zool. Syst. Evol.-forsch.*, 12:102.

34 Wiley, E.O. *The Theory and Practice of Phylogenetic Systematics*, p.262.

35 Mayr, E. *Principles of Systematic Zoology*, p.72.

36 Bruce, E.J. & Ayala, F.J. (1979) 'Phylogenetic relationships between man and the apes: electrophoretic evidence' in *Evolution*, 33:1040–56.

37 See also Goodman, M. & Moore, G.W. (1971) 'Immunodiffusion systematics of the primates I. The Catarrhini' in *Syst. Zool.*, **20**:19–62; Kohne, D.E. Chicson, J.A. & Hoyer, B.H. (1972) 'Evolution of primate DNA sequences' in *J. Human Evol.*, 1:627–44; King, M.C. & Wilson, A.C. (1975) 'Evolution at two levels in humans and chimpanzees' in *Science*, **188**:107–16. King and Wilson stipulate that the average human polypeptide is more than 99% identical to its chimpanzee counterpart.

38 Gorman, G.C., Wilson, A.C. & Nakanishi, M. (1971) 'A biochemical approach towards the study of reptilian phylogeny. Evolution of serum albumin and lactic dehydrogenase' in *Syst. Zool.*, 20:167–85.

39 Patterson, C. (1981) in *Ann. Rev. Ecol. & Syst.*, 12:195–223.

40 *Ibid.*, pp.198–9. Reproduced, with permission, from the Annual Review of Ecology & Systematics, Vol. 12 © 1981 by Annual Reviews Inc.

41 Included within Mayr's criteria for the determination of taxonomic 'weight' of characters are complexity, joint possession of derived characters, constancy through large groups, characters not associated with an 'ad hoc' specialisation, characters not affected by ecological shifts, and correlated suites of characters.

42 Mayr, E. (1974) in *Z. Zool. Syst. Evol.-forsch*, 12:466.

43 An excellent example can be seen in the North American Salamander *Plethodon*. This genus has existed for around 80 million years, and is divided into 26 species on the basis of geographical distribution, adult body size, and similarities in

proteins. But there is *no* difference in form; *Plethodon* is a morphologically monotonous group with respect to anatomy and osteology. It is possible, however, to divide this group into species on the basis of genome size. Speciation has been accompanied by a dramatic growth in the amount of DNA per haploid chromosome. This DNA increase has proceeded proportionally so that relative lengths, centromere indices and arm ratios of corresponding chromosomes are the same. There has been unimpeded expansion of the genome during speciation. See Macgregor, H.C. (1982) 'Big chromosomes and speciation amongst Amphibia' in *Genome Evolution. Syst. Assoc. Special Publ., Vol. 20*, eds. G.A. Dover & R.B. Flavell. London: Academic Press.

44 With Mayr's emphasis on classifications as theories, it might have been expected that inferred genetic similarity functioned as a theoretical predicate. Such a view is precluded by the realisation that inferred genetic similarity is an empirical indicator.

45 Mayr, E. *Principles of Systematic Zoology*, p.229.

46 Mayr, E. (1974) in *Z. Zool. Syst. Evol.-forsch*, **12**:107–8.

47 Bock, W.J. in *Major Patterns in Vertebrate Evolution*, p.872.

48 Jardine, N. & Sibson, R. *Mathematical Taxonomy*, p.139.

49 A morphological gap may be characterised as a difference in one or several characters between two or more taxa.

50 Ashlock, P.D. (1979) in *Syst. Zool.*, **28**:441–50.

51 Such a strategy is aimed at decreasing the recognition of needless monotypic taxa.

52 Mayr, E. *Principles of Systematic Zoology*, p.92.

53 Phyletic gradualism holds that new species arise from the slow and steady transformation of entire populations.

54 Michener, C.D. (1970) 'Diverse approaches to systematics' in *Evolutionary Biology*, IV, eds. Th. Dobzhansky, M.K. Hecht & M. Steere. New York: Appleton-Century-Crofts.

55 Wiley, E.O. *The Theory and Practice of Phylogenetic Systematics*.

56 A typical example of microsaltation would be Eldredge & Gould's theory of punctuated equilibria, which is a direct extension of Mayr's 'peripheral isolate' model. (See Eldredge, N. & Gould, S.J. (1972) 'Punctuated equilibria: an alternative to phyletic gradualism' in *Models in Paleobiology*, ed. T.J.M. Schopf, San Francisco: Freeman, Cooper & Co.; Gould, S.J. & Eldredge, N. (1977) 'Punctuated equilibria: the tempo and mode of evolution reconsidered' in *Paleobiology*, **3**:115–51.) Allopatric speciation (in which the diverging species occupy differing locations in space) in small, peripheral populations automatically results in gaps in the fossil record – intermediate forms of the peripheral isolate are not 'captured' (Gould, S.J. (1982) 'Punctuated equilibria – a different way of seeing' in *New Scientist*, **94**:137–41). Evolution is not a gradual process but an equilibrium punctuated by rapid change at branch points. There are long periods of stasis followed by geologically instantaneous speciation (Fig. 8). It is important to note that punctuated equilibria is not a theory about the process of macroevolution, but is concerned with large scale patterns of change – the geometry of speciation in geological time (Løvtrup, S. (1981) 'Macroevolu-

tion and punctuated equilibria' in *Syst. Zool.*, **30**:498–500). Other possible examples of microsaltation include Mayr's 'genetic revolutions' (Mayr, E. (1963) *Animal Species and Evolution*. Cambridge: Harvard Univ. Press), and Simpson's 'quantum evolution' (Simpson, G.G. *Tempo and Mode in Evolution*).

57 After Gould, S.J. (1982) in *New Scientist*, **94**:137.

58 Examples of macrosaltation are found in Goldschmidt, R.B. (1952) 'Evolution as viewed by one geneticist' in *Amer. Sci.*, **40**:84–98; Waddington, C.H. (1957) *The Strategy of the Genes*. London: Allen & Unwin; Løvtrup, S. (1974) *Epigenetics, A Treatise on Theoretical Biology*. New York: John Wiley.

59 See Thompson, P. (1983) 'Tempo and mode in evolution: punctuated equilibria and the modern synthetic theory' in *Phil. Sci.*, **50**:432–52. Thompson argues that such a view must be taken otherwise Mayr's founder principle – a mechanism of rapid non-adaptive change – would be impossible within the framework of the synthetic theory. This may be so, but Mayr's taxonomic writing is implicitly gradualistic.

60 Mickevich, M.F. (1978) 'Taxonomic congruence' in *Syst. Zool.*, **27**:143–58.

61 Mayr, E. *Principles of Systematic Zoology*, p.78.

62 Putnam, H. (1975) *Mind, Language and Reality, Philosophical Papers*, Vol. 2. Cambridge Univ. Press.

63 Kripke, S. (1972) 'Naming and necessity' in *Semantics of Natural Languages*, eds. G. Harman & D. Davidson. Dordrecht: D. Reidel.

64 Wiggins, D. (1980) *Sameness and Substance*. Oxford: Blackwell.

65 Platts, M. (1983) 'Explanatory kinds' in *Brit. J. Phil. Sci.*, **34**:133–48.

66 *Ibid.*, p.134.

67 Wiggins, D. *op. cit.*, pp.78–80 (Quoted from Platts).

68 Platts, M. *op. cit.*, p.135.

69 Mayr, E. (1975) 'The unity of the genotype' in *Biologische Zentralblatt*, **94**:381.

70 In this respect, Mayr's attempts to base his classification on the delimitation of shared ancestral genotypes, belies a clear affinity with the notion of Bauplan, than was perhaps intended.

71 Romer, A.S. & Parsons, T.S. (1977) *The Vertebrate Body*. 5th ed. Philadelphia: W.B. Saunders & Co.

72 Ehrlich, P.R. & Ehrlich, A.H. (1967) 'The phenetic relationships of butterflies' in *Syst. Zool.*, **16**:301–17.

73 I argue in *Neo-Darwinism under attack: is there any evidence of a revolution in the contemporary evolutionary debate?* (M. Phil. Thesis, Cambridge (1982)) that Mayr is, ironically enough, an unconscious founder of transformed cladistics in that he concentrates on the phenetics of certain characters in Recent organisms. It is also possible to see certain parallels in the arguments between Mayr and Simpson and the more recent debates surrounding transformed cladistics.

74 Simpson, G.G. *Tempo and Mode in Evolution*, p.xvii.

75 Newell, N.B. (1959) 'The Nature of the fossil record' in *Proc. Am. Phil. Soc.*, **103**:275.

76 Mayr, E., Linsley, E.G. & Usinger, R.L. (1953) *Methods and Principles of Systematic Zoology*. New York: McGraw-Hill, p.168.

77 Jardine, N. *Studies in the Theory of Classification*, p.126.

Chapter 3 Phylogenetic cladistics and theoretical information

1 Eldredge, N. & Cracraft, J. (1980) *Phylogenetic Patterns and the Evolutionary Process.*

2 Wiley, E.O. (1981) *The Theory and Practice of Phylogenetic Systematics.*

3 In *Principles of Animal Taxonomy* Simpson correlates horizontal relationships with relationships among contemporaneous taxa of more or less distinct common origin, while vertical relationships are seen among successive taxa of more or less distant common origin. See also Bigelow, R.S. (1961) 'Higher categories and phylogeny' in *Syst. Zool.*, **10**:86–91.

4 For a selection of references see Stensiö, E. (1968) 'The cyclostomes with special reference to the diphyletic origin of the Petromyzontida and Myxinoidea' in *Current Problems of Lower Vertebrate Phylogeny. Nobel Symposium 4*, ed. T. Ørvig. Stockholm: Almquist & Wiskell.

5 Jarvik, E. (1948) 'On the morphology and taxonomy of the middle devonian Osteolepid fishes of Scotland' in *Kungla. Svenska. Vetensk. Akad. Handlingar.*, (3) **25**:1–301.

6 Säve-Söderbergh, G. (1934) 'Some points of view concerning the evolution of the Vertebrates and the classification of this group' in *Ark.Zool.*, **26A**:1–20.

7 Baroni-Urbani, C. (1977) 'Hologenesis, phylogenetic systematics and evolution' in *Syst. Zool.*, **26**:343–6.

8 Nelson & Platnick in *Systematics and Biogeography* also cite Mitchell [see Mitchell, P.C. (1901) 'On the intestinal tract of birds; with remarks on the valuation and nomenclature of zoological characters' in *Trans. Linn. Soc. Lond.*, *Ser.2*, 8:173–275] as a forerunner of Hennig in that he anticipated the concept of shared derived characters and the notion of a 'phyletic tree – one which requires no chopping during the conversion into a classification' (Sensu Nelson & Platnick).

9 Hennig's view of systematics is more broad than that outlined in the *Introduction*, in that it is '. . . the creation of a general reference system and the pursuit of the relationship between this and other possible systems', (*Phylogenetic Systematics*: All quotes taken 2nd edn., trans. D. Dwight Davis & R. Zangerl (1979), p.8). Hennig was opposed to the view that systemtics required the creation of all possible systems that can be based on the relations among organisms, or that taxonomy involved ordering without saying anything about the way in which this order came into being. Thus the phylogenetic system must be the general reference system of systematics.

10 The derivation here is not to be confused with the mediaeval usage of the term 'clade', meaning a disaster, plague or calamity, however appropriate opponents of cladistics may find this.

11 Hennig himself uses the term 'clade' in two different ways – in the sense in which it is approximately equivalent to 'phylum' and in the terminology of Huxley where it signifies a 'delimitable monophyletic unit' (in (1958) *Uppsala Univ. Arssk.*, **6**:21–39).

12 Wiley, E.O. *The Theory and Practice of Phylogenetic Systematics.*

13 Cladograms are branching diagrams of entities and are based on inferred historical connections between the entities. They are historical dendrograms.

14 By and large, Hennig's terminology for characters and their states is confusing. Nevertheless, his concept of synapomorphy is central to all phylogenetic cladistic methods and must therefore be included.

15 Eldredge, N. & Cracraft, J. *Phylogenetic Patterns and the Evolutionary Process*, p.33.
16 I follow Eldredge & Cracraft's use of the terms 'character' and 'character state' here. These terms merely refer to similarities at two different hierarchic levels.
17 Hennig, W. *Phylogenetic Systematics*, p.73.
18 As has already been noted, definitions of strict monophyly inhibit the inclusion of fossils.
19 Eldredge, N. & Cracraft, J. *Phylogenetic Patterns and the Evolutionary Process*, p.158.
20 *Ibid.*, p.40.
21 After Felsenstein, J. (1982) in *Quart. Rev. Biol.*, **57**:380–1.
22 Cladistic analysis can therefore be performed using taxa of species or higher rank because synapomorphies already establish monophyly (nested sets) within a cladogram.
23 Lewontin, R.C. (1974) *The Genetic Basis of Evolutionary Change*. New York: Columbia Univ. Press; Throckmorton, L.H. (1966) *The Relationships of the Endemic Hawaiian Drosophilidae*. Univ. Texas Publ., No. 6615.
24 Butler, P.M. 'Directions of evolution in the mammalian dentition' in *Problems of Phylogenetic Reconstruction*.
25 Halstead, L.B. in *Problems of Phylogenetic Reconstruction*.
26 See White, M.J.D. (1978) [*Modes of Speciation*. San Francisco: W.H. Freeman, p.38] who discusses examples of speciation which are accompanied by little genetic difference, e.g. in recently separated species differences may lie in relative allele frequencies.
27 Farris, J.S. (1970) 'Methods for computing Wagner trees' in *Syst. Zool.*, **19**:83–92.
28 Wilson, E.O. (1965) 'A consistency test for phylogenies based on contemporaneous species' in *Syst. Zool.*, **14**:214–20.
29 LeQuesne, W.J. (1969) 'A method of selection of characters in numerical taxonomy' in *Syst. Zool.*, **18**:201–5.
30 Estabrook, G.F. (1972) 'Cladistic methodology: a discussion of the theoretical basis for the induction of evolutionary history' in *Ann. Rev. Ecol. & Syst.*, 3:427–56.
31 Estabrook, G.F., Johnson, C.S. & McMorris, F.R. (1975) 'An idealised concept of the true cladistic character' in *Math. Biosc.*, **23**:263–72.
32 Felsenstein, J. (1982) in *Quart. Rev. Biol.*, **57**:390.
33 Bishop, M.J. & Friday, A.E. (1982) 'Evolutionary processes: two deterministic models and their implications'. Unpublished manuscript.
34 Eldredge, N. & Cracraft, J. *Phylogenetic Patterns and the Evolutionary Process*, p.50.
35 *Ibid.*, p.64, Fig. 2.14.
36 *Ibid.*, p.68, Fig. 2.15.
37 Wiley, E.O. *The Theory and Practice of Phylogenetic Systematics*, p.97.
38 Eldredge, N. & Cracraft, J. *Phylogenetic Patterns and the Evolutionary Process*, p.129.
39 Nelson, G.J. & Platnick, N.I. *Systematics and Biogeography*, p.171.
40 Hull, D.L. (1979) 'The limits of cladism' in *Syst. Zool.*, **28**:427.
41 See Crowson, R.A. (1970) *Classification and Biology*. Atherton Press, p.266; Ashlock, P.D. (1974) 'The uses of cladistics' in *Ann. Rev. Ecol. & Syst.*, **5**: 81–99; Bock, W.J. in *Major Patterns in Vertebrate Evolution*.

42 McKenna, M.C. (1975) 'Towards a phylogenetic classification of the Mammalia' in *Phylogeny of the Primates*, eds. W.P. Luckett & F.S. Szalay. New York: Plenum Press.

43 Sister groups are defined as coordinate subsets of any set defined by a synapomorphy.

44 Nelson, G.J. (1972) 'Phylogenetic relationship and classification' in *Syst. Zool.*, **21**:227–30; also 'Classification as an expression of phylogenetic relationships' (1973) in *Syst. Zool.*, **22**:344–59.

45 Patterson, C. & Rosen, D.E. (1977) 'Review of ichthyodectiform and other Mesozoic teleost fishes and the theory and practice of classifying fossils' in *Bull. Amer. Mus. Nat. Hist.*, **158**:81–172.

46 Eldredge, N. & Cracraft, J. *Phylogenetic Patterns and the Evolutionary Process*, pp.218–19.

47 Hennig, W. *Phylogenetic Systematics*, p.210.

48 *Ibid.*, p.211.

49 In plant phylogenies this is an unrealistic assumption.

50 Platnick, N.I. & Nelson, G.J. (1978) 'A method of analysis for historical biogeography' in *Syst. Zool.*, **27**:10.

51 Hennig, W. *Phylogenetic Systematics*, p.210.

52 Wiley, E.O. (1978) 'The evolutionary species concept reconsidered' in *Syst. Zool.*, **27**:22; also in his *The Theory and Practice of Phylogenetic Systematics*.

53 Hull, D.L. (1979) in *Syst. Zool.*, **28**:416–40.

54 *Ibid.*, p.419.

55 Nelson, G.J. (1973) 'Monophyly again? : a reply to P.D. Ashlock' in *Syst. Zool.*, **22**:311.

56 Nelson, G.J. (1972) in *Syst. Zool.*, **21**:227.

57 Engelmann, G.F. & Wiley, E.O. (1977) 'The place of ancestor–descendant relationships in phylogeny' in *Syst. Zool.*, **26**:1–11.

58 After Arnold, E.N. (1981) 'Estimating phylogenies at low taxonomic levels' in *Z. Zool. Syst. Evol.-forsch.*, **19**:1–35.

59 Simpson, G.G. *Principles of Systematic Zoology*.

60 Dobzhansky, Th., Ayala, F.J., Stebbins, G.L. & Valentine, J.W. *Evolution*, p.328.

61 Platnick, N.I. (1977) 'Review of concepts of species' in *Syst. Zool.*, **26**:97.

62 Hennig, W. *Phylogenetic Systematics*, p.65.

63 Hennig implicitly argues for species as individuals.

64 Hull, D.L. (1979) in *Syst. Zool.*, **28**:431.

65 See Ghiselin, M.T. (1966) 'On psychologism in the logic of taxonomic principles' in *Syst. Zool.*, **15**:207–15; also 'A radical solution to the species problem' in *Syst. Zool.*, **23**:536–44 (1974); Hull, D.L. (1976) 'Are species really individuals?' in *Syst. Zool.*, **25**:174–91; also (1978) 'A matter of individuality' in *Phil. Sci.*, **45**:335–60; Kitts, D.B. & Kitts, D.J. (1979) 'Biological species as natural kinds' in *Phil. Sci.*, **46**:613–22; Caplan, A.L. (1980) 'Have species become declassé?' in *Philosophy of Science Association*, Vol. 1, eds. P.A. Asquith & R.N. Giere. Michigan Philosophy of Science Association; Caplan, A.L. (1981) 'Back to class: A note on the ontology of species' in *Phil. Sci.*, **48**:130–40; Wiley,

E.O. (1980) 'Is the evolutionary species fiction? – a consideration of classes, individuals and historical entities' in *Syst. Zool.*, **29**:76–80.

66 Hull, D.L. (1978) in *Phil. Sci.*, **45**:336.

67 Hull, D.L. (1979) in *Syst. Zool.*, **28**:432.

68 Dummett, M. (1973) *Frege: Philosophy of Language*. London: Gerald Duckworth & Co., p.491.

69 Bonde, N. in *Major Patterns in Vertebrate Evolution*, p.754.

70 Wiley, E.O. (1978) in *Syst. Zool.*, **27**:21–2.

71 *Ibid.*, p.22.

72 Mayr, E. *Animal Species and Evolution*.

73 Carson, H.L. (1970) 'Chromosome tracers of the origins of species: some Hawaiian *Drosophila* species have arisen from single founder individuals in less than a million years' in *Science*, **168**:1414–18.

74 Hennig, W. *Phylogenetic Systematics*, p.207.

75 Brundin, L. 'Application of phylogenetic principles in systematic and evolutionary theory' in *Current Theories in Lower Vertebrate Phylogeny*, pp.473–95.

76 Schlee, D. (1971) in *Aufsätze u. Reden. Senckenberg. Naturforsch. Ges.*, **20**:1–62.

77 Nelson, G.J. (1971) 'Cladism as a philosophy of classification' in *Syst. Zool.*, **20**:373–6.

78 Hennig, W. *Phylogenetic Systematics*, pp.56–65.

79 Hennig, W. *Phylogenetic Systematics*; Griffiths, G.C.D. (1976) 'The future of Linnaean nomenclature' in *Syst. Zool.*, **25**:168–73.

80 For instance Løvtrup, S. (1973) 'Classification, convention and logic' in *Zool. Scripta*, **2**:49–61; Nelson, G.J. (1972) in *Syst. Zool.*, **21**:227–30; also (1973) in *Syst. Zool.*, **22**:344–59; Patterson, C. & Rosen, D.E. (1977) in *Bull. Am. Mus. Nat. Hist.*, **158**:81–172; Wiley, E.O. (1979) 'An annotated Linnaean hierarchy with comments on natural taxa and competing systems' in *Syst. Zool.*, **28**:308–37.

81 Swainson, W. (1834) *A Preliminary Discourse on the Study of Natural History*. Longman.

82 They are close to an ultrametric.

83 Eldredge, N. & Cracraft, J. *Phylogenetic Patterns and the Evolutionary Process*, p.211.

84 Gregg, J.R. (1954) *The Language of Taxonomy*. New York: Columbia Univ. Press.

85 Hennig, W. *Phylogenetic Systematics*, pp.16–21.

86 *Ibid.*, pp.17, 19.

87 *Ibid.*, p.17.

88 Beckner, M. *The Biological Way of Thought*.

89 Gregg, J.R. (1967) *op. cit.*; also 'Finite Linnaean structures' (1967) in *Bull. Math. Biophys.*, **29**:191–266.

90 Buck, R.C. & Hull, D.L. (1966) 'The logical structure of the Linnaean hierarchy' in *Syst. Zool.*, **15**:97–111.

91 Jardine, N. (1969) in *Syst. Zool.*, **18**:37–52; Jardine, N. & Sibson, R. *Mathematical Taxonomy*.

92 Jardine, N. (1969) in *Syst. Zool.*, **18**:46.

93 Griffiths, G.C.D. (1974) 'On the foundations of biological systematics' in *Acta Biotheoretica*, **23**:85–131.
94 Forey, P.L. in *Problems of Phylogenetic Reconstruction*.
95 Patterson, C. (1981) in *Ann. Rev. Ecol. & Syst.*, **12**:195–223.
96 *Ibid.*, p.197.
97 *Ibid.*, p.221.
98 Patterson, C. 'The contribution of paleontology to teleostean phylogeny' in *Major Patterns in Vertebrate Evolution*.
99 Patterson, C. (1980) 'Origins of tetrapods: historical introduction to the problem' in *The Terrestrial Environment and the Origin of Land Vertebrates*, ed. A.L. Panchen. London: Academic Press.
100 *Ibid.*, p.159.
101 Ghiselin, M.T. (1969) *The Triumph of the Darwinian Method*. Berkeley: Univ. California Press, p.83.
102 Rosen, D., Forey, P., Gardiner, M. & Patterson, C. (1981) 'Lungfishes, tetrapods, paleontology, and plesiomorphy' in *Bull. Am. Mus. Nat. Hist.*, **167** (4), 159–276.
103 Patterson, C. (1981) in *Ann. Rev. Ecol. & Syst.*, **12**:195.
104 Bretsky, S.S. 'Recognition of ancestor–descendant relationships in vertebrate paleontology' in *Phylogenetic Analysis and Paleontology*.
105 Gingerich, P. in *Phylogenetic Analysis and Paleontology*, p.73.
106 *Ibid.*, p.55.
107 Paul has recently argued that the fossil record is much more complete than previously thought. See Paul, C.R.C. 'The adequacy of the fossil record' in *Problems of Phylogenetic Reconstruction*.
108 Schaeffer, B., Hecht, M.K. & Eldredge, N. (1972) 'Phylogeny and paleontology' in *Evolutionary Biology*, **6**:31–46; Bishop, M.J. (1982) 'Criteria for the determination of character state changes' in *Zool. J. Linn. Soc.*, **74**:197–206.
109 It is important to realise that the fit of an additive model tree to the data distances is formally independent of the location of the root. The best fitted tree is undirected. For conversion into classifications, trees have to be rooted.
110 Eldredge, N. & Cracraft, J. *Phylogenetic Patterns and the Evolutionary Process*, p.168.
111 *Ibid.*, p.169.
112 Farris, J.S. in *Major Patterns in Vertebrate Evolution*.

Chapter 4 Phenetics and the descriptive attitude
1 See Gilmour, J.S.L. (1937) 'A taxonomic problem' in *Nature*, **139**:1040–2. Also Gilmour, J.S.L. 'Taxonomy and philosophy' in *The New Systematics*, ed. J.S. Huxley (1940), pp.461–74, and Gilmour, J.S.L. & Walters, S.M. (1963) 'Philosophy and classification' in *Vistas in Botany*, Vol. 4:1–22, ed. W.B. Turrill. London: Pergamon Press.
2 Sneath, P.H.A. (1964) 'Mathematics and classification from Adanson to the present' in *Adanson. The Bicentennial of Michel Adanson's 'Familles des Plantes'*, ed. G.H.M. Lawrence. Pittsburg: Carnegie Institute of Technology, part 2, pp. 471–98.
3 For a full account see Guédès, M. (1967) 'La méthode Taxonomique d'Adanson' in *Rev. Hist. Sci.*, **20**:361–86.

4 For a discussion of such methods see Felsenstein, J. (1982) 'Numerical methods for inferring evolutonary trees' in *Quart. Rev. Biol.*, **57**:379–404, and Platnick, N.I. & Funk, V.A. (1982) *Advances in Cladistics*, Vol. II. Proc. Willi Hennig Soc.: New York: Columbia Univ. Press.

5 Gilmour, J.S.L. (1937) in *Nature*, **139**:1040.

6 Sneath, P.H.A. & Sokal R.R. *Numerical Taxonomy*, p.29.

7 Johnson, L.A.S. (1968) 'Rainbow's end: The quest for an optimum taxonomy' in *Proc. Linn. Soc. N.S.W.*, **93**:8–45.

8 Bridgmann, W.P. (1927) *The Logic of Modern Physics*. New York: Macmillan.

9 Sokal, R.R. & Camin, J.H. (1965) 'The two taxonomies: areas of agreement and conflict' in *Syst. Zool.*, **14**:179.

10 To talk of evolutionary theory as meaningless is too strong for what the pheneticists intended. The operational impulse was directed at the application of evolutionary theory on non-objective criteria.

11 Sneath, P.H.A. (1961) 'Recent developments in theoretical and quantitative taxonomy' in *Syst. Zool.*, **10**:124.

12 Taxonomic relationships are evaluated purely on the basis of resemblances existing *now* in the material at hand.

13 Duncan, T. & Baum, B.R. (1981) 'Numerical phenetics: its use in botanical systematics' in *Ann. Rev. Ecol. & Syst.*, **12**:387–404.

14 Sokal, R.R. (1966) 'Numerical Taxonomy' in *Scient. Am.*, **215**(6):112.

15 It must be remembered that phenetic studies are not confined to morphological characters alone as some have implied (e.g. Bonde, N. 'Cladistic classifications as applied to vertebrates' in *Major Patterns in Vertebrate Evolution*; Gingerich, P.D. (1979) 'The stratophenetic approach to phylogeny reconstruction' in *Phylogenetic Analysis and Paleontology*, eds. J. Cracraft & N. Eldredge. New York: Columbia Univ. Press); anatomical, cytological, chemical and behavioural characteristics are equally appropriate to the phenetic approach, see McNeill, J. (1979) 'Purposeful phenetics' in *Syst. Zool.*, **28**:465–82.

16 Jardine, N. & Sibson, R. *Mathematical Taxonomy*.

17 Cain, A.J. & Harrison, G.A. (1958) 'An analysis of the taxonomist's judgement of affinity' in *Proc. Zool. Soc. Lond.*, **131**:85–98.

18 Sokal, R.R. & Rohlf, F.J. (1970) 'The intelligent ignoramus: an experiment in numerical taxonomy' in *Taxon*, **19**:305–19.

19 Dunn, G. & Everitt, B.S. *An Introduction to Mathematical Taxonomy*.

20 Sneath, P.H.A. & Sokal, R.R. *Numerical Taxonomy*.

21 *Ibid.*, p.22.

22 Jardine, N. (1971) 'Patterns of differentiation between human local populations' in *Phil. Trans. Roy. Soc. Lond. B*, **263**:13.

23 The distinction between agglomerative techniques – those which work by successive fusion of the *n* OTUs into groups – and divisive techniques – which function by separating the set of *n* OTUs simultaneously (Dunn, G. & Everitt, B.S. *An Introduction to Mathematical Taxonomy*, p.77; Sneath, P.H.A. & Sokal, R.R. *Numerical Taxonomy*, p.203) must not be confused with methods in phenetics. These techniques are algorithms for data processing. Hierarchic stratified cluster methods, e.g. average link methods, can be done agglomeratively or divisively. Hence it is important to distinguish between methods and algorithms in phenetics.

24 Jardine, N., Van Rijsbergen, C.J. & Jardine, C.J. (1969) 'Evolutionary rates and the inference of evolutionary tree forms' in *Nature*, 244:185.

25 After Schnell, G.D. (1970) 'A phenetic study of the suborder Lari (Aves)' in *Syst. Zool.*, 19:264–302; fig. 12, p.266. Reproduced from Nelson, G.J. & Platnick, N.I. *Systematics and Biogeography*, p.277.

26 Farris, J.S. 'On the phenetic approach to vertebrate classification' in *Major Patterns in Vertebrate Evolution*.

27 *Ibid.*, p.829.

28 Cain, A.J. (1954) *Animal Species and Evolution*. London: Hutchinson; Cain, A.J. & Harrison, G.A. (1958) 'An analysis of the taxonomist's judgement of affinity' in *Proc. Zool. Soc. Lond.*, 131:85–98; Cain, A.J. & Harrison, G.A. in *Proc. Zool. Soc. Lond.*, 135:1–31.

29 Gilmour, J.S.L. (1961) in *Nature*, 139:1040–2; also in *The New Systematics*, and 'Taxonomy' in *Contemporary Botanical Thought*, eds. A.M. MacCleod & C.S. Cobley. Edinburgh: Oliver & Boyd.

30 Sokal, R.R. & Sneath, P.H.A. *Principles of Numerical Taxonomy*.

31 Sneath, P.H.A. & Sokal, R.R. *Numerical Taxonomy*.

32 Reference is made to Michener, C.D. (1963) 'Some future developments in taxonomy' in *Syst. Zool.*, 12:151–72.

33 Mill, J.S. (1974) 'A synthesis of logic ratiocinative and inductive. Books I–VIII: IV: Of operations subsidiary to induction' in *Collected Works of John Stuart Mill*, Vol. IV, ed. J. Robson. Toronto: Univ. Toronto Press, p.714.

34 For instance, from the class of 'sharp things' and the class of 'things that hurt', we can inductively infer that 'sharp things hurt'.

35 Gilmour, J.S.L. (1937) in *Nature*, 139:1042.

36 Sokal & Sneath prefer to use the term 'character-state' instead of 'attribute'. For them a character is 'any feature which varies from one individual organism to another' (*Principles of Numerical Taxonomy*, p.61) while a character-state is equivalent to 'any attribute of a member of a taxon by which it differs or may differ from a member of a different taxon' (after Mayr, E. *Principles of Systematic Zoology*, p.413). For example, in beetles the condition of the elytral surface would be a character, while 'smooth' or 'punctuate' would be character-states.

37 Gilmour, J.S.L. & Walters, S.M. in *Vistas in Botany*, IV, p.3.

38 *Ibid.*, p.4.

39 *Ibid.*

40 Gilmour, J.S.L. in *Contemporary Botanical Thought*, p.33.

41 As in the problem of convergence.

42 Sneath, P.H.A. & Sokal, R.R. *Numerical Taxonomy*, p.5.

43 Jardine, N. & Sibson, R. *Mathematical Taxonomy*, p.137.

44 Sneath, P.H.A. & Sokal, R.R. *Numerical Taxonomy*, p.17.

45 This is based on the observation that phenetic methods lead to inconsistent groupings in theory [Farris, J.S. in *Major Patterns in Vertebrate Evolution*] and practice [Mickevich, M.F. (1978) in *Syst. Zool.*, 27:143–58; Nelson, G.J. (1979) in *Syst. Zool.*, 28:1–21.] Rohlf & Sokal (in 'Comments on taxonomic congruence' in *Syst. Zool.*, 29:97–101 (1980)) subsequently replied that Mickevich's analysis was biased because it employed cladistic techniques and assumptions.

46 Michener, C.D. (1973) in *Syst. Zool.*, **12**:151–72; Ruse, M. (1973) *The Philosophy of Biology*. London: Hutchinson & Co.

47 Farris, J.S. (1982) in *Syst. Zool.*, **31**:413.

48 Nor can any mathematical justification be found. For example, in linear regression, justification for least squares is based on the fact that each set of y points is *normally* distributed. In phenetic methods of best fit no equivalent can be found.

49 Eldredge, N. & Cracraft, J. *Phylogenetic Patterns and the Evolutionary Process.*

50 Sneath, P.H.A. & Sokal, R.R. *Numerical Taxonomy*, p.69.

51 *Ibid.*

52 Sokal, R.R. & Sneath, P.H.A. *Principles of Numerical Taxonomy*, p.161.

53 Cain, A.J. & Harrison, G.A. (1958) in *Proc. Zool. Soc. Lond.*, **131**:85–98.

54 Cain, A.J. & Harrison, G.A. (1960) in *Proc. Zool. Soc. Lond.*, **135**:3.

55 *Ibid.*

56 Cain, A.J. & Harrison, G.A. (1958) in *Proc. Zool. Soc. Lond.*, **131**:90.

57 This would exclude sets of characters controlled by single genes.

58 Moss, W.W. (1972) 'Some levels of phenetics' in *Syst. Zool.*, **21**:236–9.

59 Johnson, L.A.S. (1968) in *Proc. Linn. Soc. N.S.W.*, **93**:8–45.

60 Dobzhansky, Th. (1970) *The Genetics of the Evolutionary Process*. New York: Columbia Univ. Press, p.312.

61 Cain, A.J. & Harrison, G.A. (1960) in *Proc. Zool. Soc. Lond.*, **135**:1–31; Sokal, R.R. & Sneath, P.H.A. *Principles of Numerical Taxonomy.*

62 Weak phenetics is also guilty of this.

63 Dupraw, E.J. (1964) 'Non-Linnaean taxonomy' in *Nature*, **202**:849–52; (1965) 'Non-Linnaean taxonomy and the systematics of honeybees' in *Syst. Zool.*, **14**:1–24; (1965) 'The recognition and handling of honeybee specimens in non-Linnaean taxonomy' in *J. Apicult. Res.*, **4**:71–84.

64 Recently, non-hierarchical systems have been used at infraspecific levels, leading to the view that hierarchic systems are suitable for supraspecific classifications, while non-hierarchical systems are suitable for infraspecific classifications (Jardine, N. & Sibson, R. *Mathematical Taxonomy*).

65 Jardine, N. (1969) in *Syst. Zool.*, **18**:59 *(fn)*.

66 Dupraw, E.J. (1964) in *Nature*, **202**:849.

67 Each non-Linnaean classification consists of a two-dimensional distribution of specimen points, and axes of the distribution being defined in each case by two sets of discriminant coefficients. Specimen points are plotted on a multidimensional graph, the axes of which are the scales of the specimen variables. Specimen points in the character hyperspace exhibit clustering.

68 Dupraw, E.J. (1965) in *J. Apicult. Res.*, **4**:73.

69 Sokal, R.R. & Sneath, P.H.A. *Principles of Numerical Taxonomy*, p.191.

70 Sokal, R.R. (1962) 'Typology and empiricism in taxonomy' in *J. Theor. Biol.*, **3**:246.

71 Jardine, N. & Sibson, R. *Mathematical Taxonomy*; Jardine, N. (1971) in *Phil. Trans. Roy. Soc. Lond.B*, **263**:1–33.

72 Jardine, N. *op. cit.*, p.13.

73 Dupraw, E.J. (1965) in *Syst. Zool.*, **14**:11.

74 Sokal, R.R. & Rohlf, F.J. (1970) in *Taxon*, **19**:305–19.

75 Gilmour, J.S.L. in *The New Systematics*.

204 Notes to chapter 4

76 Dingle, H. (1938) 'The rational and empirical elements in physics' in *Philosophy*,
   13:148–65.
77 Gilmour, J.S.L. in *The New Systematics*, p.468.
78 *Ibid.*, p.464.
79 Gilmour, J.S.L. & Walters, S.M. in *Vistas in Botany*, IV.
80 Logical positivism arose in Vienna in the 1920s with the formation of the
   famous discussion group, the Vienna Circle, led by Moritz Schlick, Rudolf
   Carnap and Otto Neurath.
81 Sneath, P.H.A. & Sokal, R.R. *Numerical Taxonomy*, p.9.
82 An analogy can be seen in psychology, where psychological theory and behaviourism are the counterparts of evolutionary theory and classification.
   Psychological theory traditionally embraced the introspectionists, physiologists
   and Freudians since they all reverted to the use of theories in one form or
   other. The method of the introspectionists was to analyse consciousness itself,
   by examining either the verbal reports of a subject or one's own feelings and
   perceptions. Freud went on to build a theoretical structure on such data and
   turned his attention to therapy, while the physiologists attempted to divorce
   themselves from these two schools by concentrating on the central nervous
   system alone.
       It was in the midst of such factionalism that J.B. Watson introduced the
   notion of behaviourism as an extreme reaction to the introspectionism of
   Wundt and Titchener, and one of its early founders William James (see Watson,
   J.B. (1914) *Behaviour: Introduction to Comparative Psychology*. New York: Holt;
   also his (1924) *Behaviourism*. New York: Holt). Watson argued that the subject
   must be viewed as a black box since all that mattered was input and output.
   The task of the behaviourist was simple: 'His sole object is to gather facts
   about behaviour – verify his data – subject them both to logic and mathematics
   (the tools of every scientist)' (in *Behaviourism*, p.6). In short, Watson wanted to
   purge psychology of any subjective, mental operations, with the result that all
   terms, e.g. 'thirst', 'intelligence', must be defined in terms of trembling, facial
   expressions and the like. Only operational definitions could be used. The subject is not permitted to state 'I'm afraid' since the apparent, introspective character of such utterances should be avoided; they are meaningless noises. One
   unfortunate consequence of this is that the behaviourist must confront the
   problem of giving a behavioural account of language (Skinner, B.F. (1957)
   Verbal Behaviour. New York: Appleton-Century-Crofts).
       The extreme implications in behaviourism were put forward in a classic
   paper by Skinner (1950) ('Are theories of learning necessary?' in *Psychol. Rev.*,
   57:193–216) who argued that theories, whether of a conceptual, neural or mental
   type were not necessary. Three reasons were given: (*a*) they create new problems of explanation which get covered up; (*b*) they may generate wasteful
   research; (*c*) more direct approaches exist.
       Clearly, such an extreme position in psychology is untenable; facts do not
   speak for themselves. Broadbent has argued that Skinner's approach, which
   seeks only to predict and not to explain behaviour, cannot be theory-free on
   the grounds that no two events are identical (Broadbent, D.E. (1961) *Behaviour*.
   London: Eyre & Spottiswoode). Theoretical assumptions must be invoked to

enable selection of the relevant dimensions on which generalisations can be based. For a full discussion on behaviourism see Ions, E. (1977) *Against Behaviourism. A Critique of Behavioural Science.* Oxford: Basil Blackwell; Valentine, E.R. (1982) *Conceptual Issues in Psychology.* London: Allen & Unwin.

83 Hull, D.L. (1968) in *Syst. Zool.*, **16**:438–57.

84 Although strong phenetics is unstable in the light of the fossil record, this is not a problem as in cladistics, because no knowledge claims are being made as such. Strong pheneticist arguments against fossils are largely practical. Indeed, there would be no argument were a *complete* record available.

85 On a much more general note Atran has argued that the notion of classifications serving a function is part of a long-standing empiricist–inductivist tradition. Under this view, 'progress in the development of taxonomy is marked by a move from culturally parochial "special purpose" orderings to those increasingly transcultural and general purpose' (Atran, S. (1983) 'Covert fragmenta and the origins of the botanical family' in *Man*, **18**:54). Strong pheneticists clearly adopt this view in their suggestion that Adanson's method was the only means of obtaining a natural classification, while all other classifications, e.g. those of Linnaeus, were artificial in the sense of arbitrarily favouring one character over another, and hence special purpose classifications.

Atran goes on to relate this view of taxonomic development with present-day attitudes to folk taxonomy, where all classification is subject to the purposeful dictates of symbolism and economic needs, e.g. see Berlin, B., Breedlove, D.E. & Raven, P.H. (1966) 'Folk taxonomies and biological classification' in *Science*, **154**:273–5; Mayr also takes this view in (1982) *The Growth of Biological Thought.* Cambridge, Mass.: Harvard Univ. Press, p.134. This view presupposes that there is a universal anthropocentric bias which is purposeful in the sense of being inextricably world bound. Against this view, Atran argues that there is no continuum from 'primitive' cultural symbolism to the 'adult' scientist via some changing operation, such as induction, or increasing the general purpose and de-emphasising the artificial. Instead, an innate, grounded common-sense is both prior to, and necessary for, any symbolic or scientific elaboration of the world. In folk taxonomy, every domain, e.g. artefacts, colours, people, etc., will involve very different modalities, and there are no interesting properties which cut across these domains. No classificatory modality cuts across these domains. Instead, there must be a commonsense understanding of the phenomenal world, which is manifested in a basic level of groupings. The world is thus partitioned into mutually exclusive classes, and other domains, e.g. diseases, cannot be fitted into the system.

This idea of commonsense universals is novel, and if true, would imply that the rational and empiricist prejudices inherent in strong phenetics cut much deeper than I have made out. Such prejudices would be based on the deceptive empiricist epistemology wherein human knowledge begins with some vague symbolic or semiotic capacity which artificially extends an immediate and practical knowledge of the world to 'metaphysical truth'. The fact that Linnaeus and Adanson derived very similar groupings would tend to support the view that their talk of rationalism and empiricism is only *post hoc* peripheral justification.

86 This is reflected in the less zealous version of Sneath & Sokal's *Numerical Taxonomy*.

87 Originally, Sokal & Sneath argued that it was necessarily circular to have theoretically biased data. Thus phylogenetic conclusions should not influence
   taxonomic characters. *But* to make a preliminary arrangement on general
   resemblance, test its components for phylogenetic concordance, and then base a
   classification on the results of these tests, is not circular reasoning, but more a
   case of positive feedback. This extreme attitude was later watered down to
   imply that phenetic relationships formed the basis for phylogenetic relationships, and that the phenetic and cladistic elements in classification should be
   kept separate.
88 Sokal, R.R. & Sneath, P.H.A. *Principles of Numerical Taxonomy*, p.56.
89 Farris, J.S. (1979) in *Major Patterns in Vertebrate Evolution*; also 'The information
   content of the phylogenetic system' in *Syst. Zool.*, **28**:483–519.
90 McNeill, J. (1982) 'Phylogeny reconstruction and phenetic taxonomy' in *Zool.
   J. Linn. Soc.*, **74**:337–44.

Chapter 5 Transformed cladistics and the methodological turn
1 Platnick, N.I. (1979) in *Syst. Zool.*, **28**:537–46.
2 Nelson, G.J. & Platnick, N.I. *Systematics and Biogeography*.
3 Patterson, C. (1982) in *The Biologist*, **27**:234–40; also his 'Cladistics and classification' in *Darwin up to Date*, ed. J. Cherfas. IPC Magazines.
4 Beatty, J. (1982) 'Classes and cladists' in *Syst. Zool.*, **31**:25–34.
5 Charig, A.J. 'Systematics in Biology: a fundamental comparison of some major
  schools of thought' in *Problems of Phylogenetic Reconstruction*.
6 Ridley, M. (1986) *Evolution and Classification: the reformation of cladism*. London:
  Longman.
7 Hull, D.L. (1979) in *Syst. Zool.*, **28**:416–40.
8 Simpson, G.G. (1978) 'Variation and details of macroevolution' in *Paleobiology*,
  4:217–21.
9 Panchen, A.L. (1982) 'The use of parsimony in testing phylogenetic hypotheses'
  in *Zool. J. Linn. Soc.*, **74**:305–28.
10 Nelson, G.J. (1979) 'Cladistic analysis and synthesis; principles and definitions
   with a historical note on Adanson's *Familles des Plantes* (1763–1764)' in *Syst.
   Zool.*, **28**:1–21.
11 Platnick, N.I. (1979) in *Syst. Zool.*, **28**:545.
12 Patterson, C. in *Darwin up to Date*.
13 Platnick, N.I. & Nelson, G.J. (1981) 'The purposes of biological classification'
   in *PSA 1978*, Vol. II, eds. P.D. Asquith & I. Hacking, p.121.
14 Eldredge, N. & Cracraft, J. *Phylogenetic Patterns and the Evolutionary Process*, p.6.
15 Wiley, E.O. *The Theory and Practice of Phylogenetic Systematics*, p.78.
16 Tattershall, I. & Eldredge, N. (1977) 'Fact, theory and fantasy in human
   paleontology' in *Am. Scient.*, **65**:204–11.
17 Nelson, G.J. (1979) in *Syst. Zool.*, **28**:8.
18 An excellent example is given in Janis, C. (1982) 'Evolution of horns in
   Ungulates: ecology and paleoecology' in *Biol. Rev.*, **57**:261–318.
19 Hennig, W. *Phylogenetic Systematics*.
20 Brundin, L. (1972) 'Evolution, causal biology and classification' in *Zool. Scripta.*,
   1:107–20.

21 After (i) Hennig, W. *Phylogenetic Systematics*, p.71; (ii) Nelson, G.J. (1972) 'Comments on Hennig's *Phylogenetic Systematics* and its influence on ichthyology' in *Syst. Zool.*, **21**:364–74.

22 Traditionally, homology has been defined in terms of common ancestry; thus Simpson states that 'Homology is resemblance due to inheritance from a common ancestry' (in *Principles of Animal Taxonomy*, p.78).

23 Patterson, C. 'Morphological characters and homology' in *Problems of Phylogenetic Reconstruction*.

24 *Ibid.*, p.33.

25 Wiley, E.O. (1979) 'Cladograms and phylogenetic trees' in *Syst. Zool.*, **28**:88–92.

26 For phylogenetic cladists this necessarily involves the determination of polarity of morphoclines, as either primitive or derived.

27 Ingroup comparison is a possible alternative; see Jefferies R.P.S. (1979) 'The origin of chordates – a methodological essay' in *The Origin of Major Invertebrate Groups*, ed. M.R. House. London: Academic Press.

28 Nelson, G.J. (1978) 'Ontogeny, phylogeny and the biogenetic law' in *Syst. Zool.*, **27**:324–45.

29 *Ibid.*

30 Nelson, G.J. (1973) 'The higher level phylogeny of vertebrates' in *Syst. Zool.*, **22**:87–91.

31 Nelson, G.J. & Platnick, N.I. *Systematics and Biogeography*, p.332.

32 Patterson, C. in *The Biologist*, **27**:237.

33 This is clearly wrong because the transformed cladists have attempted to present a coherent view of systematics, based on the idea that the truth of classification presupposes the truth of evolution. This is at odds with the more mainstream attitudes of the phylogenetic cladists. Furthermore, it is to some extent irrelevant whether or not transformed cladists recognise themselves as such because scientific discourse relies on the simplification and hardening of attitudes so that important issues can be brought into sharper relief for the purposes of criticism (See Hull, D.L. (1983) 'Darwin and the nature of science' in *Evolution from Molecules to Man*, ed. D.S. Bendall. Cambridge Univ. Press).

34 It should not be surprising that transformed cladistics cannot be fitted into a single category, since it is common at the forefront of research to have imprecise assumptions. Not only are such assumptions difficult to spell out, but also advances in taxonomy depend upon this lack of clarity. This is certainly the case with the shifting notions of 'cladogram', 'synapomorphy', 'pattern' and the like. Furthermore, formalistic definitions are often used to cover up imprecision; for example, Nelson & Platnick (in *Systematics and Biogeography*, p.12) define species as 'the smallest detected samples of self-perpetuating organisms that have unique sets of characters.' Since the definition utilised is always likely to be narrower in scope than the concept allegedly defined, the adoption of formalistic definitions will 'cloud over' any underlying assumptions that may be present.

35 Beatty, J. (1982) in *Syst. Zool.*, **31**:25–34; Charig, A.J. in *Problems of Phylogenetic Reconstruction*; Ridley, M. (1983) 'Can classifications do without evolution?' in *New Scientist*, **100**:647–51.

36 Brady, R.H. (1982) 'Theoretical issues and pattern cladistics' in *Syst. Zool.*, **31**:288.

37 Nelson, G.J. & Platnick, N.I. *Systematics and Biogeography*, p.324.

38 Platnick, N.I. (1982) 'Defining characters and evolutionary groups' in *Syst. Zool.*, **31**:282–4.

39 There are obvious parallels with the views of Thom and D'Arcy Thompson, and to quote the latter: 'The physicist explains interims of the properties of matter, and classifies according to a mathematical analysis . . . and his task ends there. But when such forms, such confirmations and configurations, occur among *living* things, then at once the biologist introduces his concept of heredity, of historical evolution, of suspension in time . . . of common origins of similar forms remotely separated by geographic space or geologic time, of fitness for a function, of adaptation to the environment, of higher and lower. . . This is the fundamental difference between the explanation of the physicist and those of the biologist.' (in *Growth and Form*. Quoted from Leith, B. (1982) *The Descent of Darwin*. London: Collins, p.108).

40 Patterson, C. in *Darwin up to Date*, p.36.

41 Nelson, G.J. & Platnick, N.I. *Systematics and Biogeography*, p.16.

42 Patterson, C. *op. cit.*, p.37.

43 This characterisation is based on Platnick, N.I. (1979) in *Syst. Zool.*, **28**:537–46; Brooks, D.R. (1982) 'Review of *Systematics and Biogeography* ' in *Syst. Zool.*, **31**:206–8.

44 Nelson, G.J. & Platnick, N.I. *Systematics and Biogeography*, p.5.

45 Croizat, L. (1964) *Space, Time, Form: The Biological Synthesis*. Caracas: Published by the author.

46 For example, Nelson & Platnick use phenograms disguised as cladograms on pp.495 (fig. 8.21), 497 (fig. 8.23) and 498 (fig. 8.24).

47 Farris, J.S. in *Major Patterns in Vertebrate Evolution*.

48 Mickevich, M.F. (1978) in *Syst. Zool.*, **27**:143–58.

49 Lockhart, W.R. & Hartmann, P.A. (1963) 'Formation of monothetic groups in quantitative bacterial taxonomy' in *J. Bacteriol.*, **85**:68–77.

50 Lance, G.N. & Williams, W.T. (1965) 'Computer programs for monothetic classification (association analysis)' in *Comput. J.*, **8**:246–9.

51 *Ibid.*

52 Sokal, R.R. & Sneath, P.H.A. *Principles of Animal Taxonomy*, p.14.

53 Jardine, N. & Sibson, R. *Mathematical Taxonomy*, p.115.

54 Nelson, G.J. (1979) in *Syst. Zool.*, **28**:1–21.

55 Platnick, N.I. & Nelson, G.J., in *Philosophy of Science Association 1978*, Vol. II.

56 Hennig, W. *Phylogenetic Systematics*, p.7.

57 Nelson, G.J. & Platnick, N.I. *Systematics and Biogeography*, p.23.

58 Sokal, R.R. & Sneath, P.H.A. *Principles of Numerical Taxonomy*, p.85.

59 *Ibid.*, p.64.

60 Sneath, P.H.A. & Sokal, R.R. *Numerical Taxonomy*, p.97.

61 Platnick, N.I. (1979) in *Syst. Zool.*, **28**:544.

62 Platnick, N.I. & Nelson, G.J. in *Philosophy of Science Association 1978*, Vol. II, p.119.

63 Nelson, G.J. & Platnick, N.I. *Systematics and Biogeography*, p.311.

64 *Ibid.*, pp.35–6.

65 *Ibid.*, p.9.

66 Platnick, N.I. & Nelson, G.J. *Philosophy of Science Association 1978*, Vol. II, p.123.
67 Patterson, C. in *Problems of Phylogenetic Reconstruction*, p.59.
68 Nelson, G.J. & Platnick, N.I. *Systematics and Biogeography*, p.249.
69 Ball, I.R. (1983) 'On groups, existence and the ordering of nature' in *Syst. Zool.*, **32**:446–51.
70 Mayr, E. *The Growth of Biological Thought*, p.38.
71 Hare, R.M. (1982) *Plato*. Oxford: Oxford Univ. Press.
72 See McNeill, J. [in *Zool. J. Linn. Soc.*, **74**:337–44 (1979)] who suggests that the analyses of transformed cladistics and phenetics are made possible by evolution, but that neither provide evolutionary trees, nor of themselves, permit phylogenetic reconstruction.
73 Kuhn, T.S. (1962) *The Structure of Scientific Revolutions. International Encyclopedia of Unified Science*, Vol. 2, No. 2. Chicago: Univ. of Chicago Press.
74 Indeed, the dishonesty of the transformed cladistic enterprise in this respect, becomes clearer when it is considered that within the philosophy of science, Popper's view is one amongst many, and is deficient. Science does not, and cannot, proceed by falsification alone (there are plenty of theories which contain anomalies); it should also be consilient (Ruse, M. (1979) 'Falsifiability, consilience and systematics' in *Syst. Zool.*, **28**:530–6). Science should explain many different, diverse areas, drawing all together and unifying under one single hypothesis, e.g. 'plate tectonics' in geology.
75 Settle, T. (1979) 'Popper on when is a science not a science?' in *Syst. Zool.*, **28**:521–9.
76 Platnick, N.I. & Gaffney, E.S. (1977) 'Systematics: a Popperian perspective' in *Syst. Zool.*, **26**:360–5.
77 For a selection of references see Kitts, D.B. (1977) 'Karl Popper: verifiability and systematic zoology' in *Syst. Zool.*, **26**:185–94; (1978) 'Theoretics and systematics: a reply to Cracraft, Nelson and Patterson' in *Syst. Zool.*, **27**:222–4 (1978); Platnick, N.I. & Gaffney, E.S. (1978) 'Evolutionary biology: a Popperian perspective' in *Syst. Zool.*, **27**:137–41; 'Systematics and the Popperian paradigm' in *Syst. Zool.*, **27**: 381–8.
78 Ruse, M. (1977) 'Karl Popper's philosophy of biology' in *Phil. Sci.*, **44**:638–61.
79 Popper, K.R. (1957) *The Poverty of Historicism*. London: Routledge & Kegan Paul, p.108.
80 Popper, K.R. (1974) 'Darwinism as a metaphysical research programme' in *The Philosophy of Karl Popper*, Vol. I, ed. P.A. Schlipp. La Salle III: Open Court.
81 Popper, K.R. (1972) *Objective Knowledge. An Evolutionary Approach*. Oxford Univ. Press, p.243.
82 Panchen, A.L. (1982) in *Zool. J. Linn. Soc.*, **74**:313.
83 Patterson, C. (1978) 'Verifiability in systematics' in *Syst. Zool.*, **27**:218–22.

Chapter 6 Transformed cladistics and evolution

1 Patterson, C. (1980) in *The Biologist*, **27**:234.
2 Brady, R.H. (1982) in *Syst. Zool.*, **31**:289.
3 Platnick, N.I. & Nelson, G.J. in *Proceedings of the Philosophy of Science Association 1978*, Vol. II, p.123.

4 Hill, C.R. & Crane, P.R. 'Evolutionary cladistics and the origin of angiosperms' in *Problems of Phylogenetic Reconstruction*, p.302.

5 Nelson, G.J. & Platnick, N.I. *Systematics and Biogeography*, p.321.

6 Jefferies, R.P.S. in *The Origin of Major Invertebrate Groups*, p.453.

7 Von Baer, K.E. (1977) *Entwicklungsgeschichte der Thiere*. Quoted from Gould, S.J. *Ontogeny and Phylogeny*. Cambridge, Mass.: Harvard Univ. Press, p.55.

8 Nelson, G.J. (1978) in *Syst. Zool.*, **27**:324–45.

9 Nelson, G.J. & Platnick, N.I. *Systematics and Biogeography*, p.343.

10 Voorzanger, B. & Van der Steen, W.J. (1982) 'New perspectives on the biogenetic law?' in *Syst. Zool.*, **31**:202–5.

11 Patterson, C. (1983) 'How does phylogeny differ from ontogeny?' in *Development and Evolution*, eds. B.C. Goodwin, N. Holder & C.C. Wylie. Cambridge Univ. Press.

12 Eldredge, N. & Cracraft, J. *Phylogenetic Patterns and the Evolutionary Process*; Wiley, E.O. *The Theory and Practice of Phylogenetic Systematics*.

13 Nelson, G.J. & Platnick, N.I. *Systematics and Biogeography*; Patterson, C. (1980) in *The Biologist*, **27**:234–40.

14 Farris, J.S. 'The logical basis of phylogenetic analysis' in *Advances in Cladistics II*.

15 Hennig, W. *Phylogenetic Systematics*.

16 Sober, E. (1975) *Simplicity*. Oxford: Clarendon Press.

17 Felsenstein has suggested three versions of this claim: (i) the most parsimonious outcome of evolution would be for it not to occur at all, i.e. for all species to be identical; (ii) each derived state to have arisen no more than once; (iii) given the goal of reaching the observed species, evolution has taken the shortest path. (Felsenstein, J. (1983) 'Parsimony in systematics: biological and statistical issues' in *Ann. Rev. Ecol. & Syst.*, **14**:313–33.) Hennig's original method is compatible with (i), while phenetic methods for minimum evolution reconstruction of trees are compatible with (iii) [See Panchen, A.L. (1982) in *Zool. J. Linn. Soc.*, **74**:305–28].

18 For example see Romero-Herrera, A.E., Lehman, H., Joysey, K.A. & Friday, A.E. (1973) 'Molecular evolution of myoglobin and the fossil record: a phylogenetic synthesis' in *Nature*, **246**:389–95; also (1978) 'On the evolution of myoglobin' in *Phil. Trans. Roy. Soc. Lond. B*, **283**:61–163.

19 Thus 'all swans are white' is a simpler hypothesis than 'all swans are white except Australasian swans which are black' because the extra information 'this is a swan' would enable the first hypothesis to answer the question 'what colour is it?' whereas we need further information of where the swan lives before the second hypothesis is able to answer that question. The hypothesis with fewer boundary conditions is therefore more informative than hypotheses requiring more boundary conditions.

20 Beatty, J. & Fink, W.L. (1979) '[Review of] *Simplicity* ' in *Syst. Zool.*, **28**:643–51.

21 *Ibid.*, p.647.

22 *Ibid.*

23 Sober, E. *Simplicity*, p.173.

24 Friday, A.E. (1982) 'Parsimony, simplicity and what actually happened' in *Zool. J. Linn. Soc.*, **74**:329–35.

25 Farris, J.S. (1977) 'Phylogenetic analysis under Dollo's law' in *Syst. Zool.*, **26**:77–88.

26 Sober, E. (1983) 'Parsimony in systematics: philosophical issues' in *Ann. Rev. Ecol. & Syst.*, **14**:335–57.
27 Friday, A.E. (1982) in *Zool. J. Linn. Soc.*, **74**:332.
28 Support in the likelihood method is precisely defined as the natural algorithm of the likelihood ratio between two hypotheses, and takes a numerical value.
29 Friday, A.E. (1982) in *Zool. J. Linn. Soc.*, **74**:329–35.
30 Farris, J.S. *Advances in Cladistics II.*
31 Farris suggests that parsimony implies the minimisation of the number of required homoplasies, and not that homoplasies are rare. But this still fails to overcome the problem.
32 Eldredge, N. & Cracraft, J. *Phylogenetic Patterns and the Evolutionary Process*; Wiley, E.O. *The Theory and Practice of Phylogenetic Systematics.*
33 See Eldredge, N. & Cracraft, J. *Phylogenetic Patterns and the Evolutionary Process*; Gaffney, E.S. in *Phylogenetic Analysis and Paleontology*; Nelson, G.J. & Platnick, N.I. *Systematics and Biogeography*; Wiley, E.O. (1975) in *Syst. Zool.*, **24**:233–43.
34 Wiley, E.O. *The Theory and Practice of Phylogenetic Systematics*, p.110.
35 Nelson, G.J. & Platnick, N.I. *Systematics and Biogeography*, p.342.
36 Popper, K.R. (1959) *The Logic of Scientific Discovery.* London: Hutchinson.
37 Panchen, A.L. (1982) in *Zool. J. Linn. Soc.*, **74**:323.
38 Patterson, C. (1980) in *The Biologist*, **27**:236.
39 Patterson, C. in *Problems of Phylogenetic Reconstruction*, p.42.
40 Patterson, C. (1980) in *The Biologist*, **27**:234–40.
41 Panchen, A.L. (1982) in *Zool. J. Linn. Soc.*, **74**:305–28.
42 Nelson, G.J. & Platnick, N.I. *Systematics and Biogeography*, p.220.
43 Van Valen, L. 'Why not to be a cladist' in *Evol. Theory*, **3**:289 (1978).
44 Patterson, C. (1978) in *Syst. Zool.*, **28**:220.
45 Panchen, A.L. Personal communication.
46 Nelson, G.J. & Platnick, N.I. *Systematics and Biogeography*, p.35; also Platnick, N.I. (1982) in *Syst. Zool.*, **31**:283.
47 Leith, B. *The Descent of Darwin*, p.97.
48 Platnick, N.I. (1982) in *Syst. Zool.*, **31**:283.
49 Patterson, C. in *The Terrestrial Environment and the Origin of Land Vertebrates.*
50 This is certainly Darwin's attitude: for him, evolution was the hypothesis which explained the supposedly real relationships evident in a natural hierarchic classification.
51 Nelson, G.J. & Platnick, N.I. *Systematics and Biogeography*, p.164.
52 Brady has also argued along similar lines (in 'Parsimony, hierarchy and biological implications' in *Advances in Cladistics II*).
53 Wiley, E.O. *The Theory and Practice of Phylogenetic Systematics*, p.71.
54 Platnick, N.I. (1979) in *Syst. Zool.*, **28**:537–46.
55 Transformed cladists do not take into account non-causal explanations.
56 Gaffney, E.S. in *Phylogenetic Analysis and Paleontology*, p.86.
57 Eldredge, N. & Cracraft, J. *Phylogenetic Patterns and the Evolutionary Process.*
58 Patterson, C. in *Darwin up to Date*, p.37.
59 Nelson, G.J. & Platnick, N.I. *Systematics and Biogeography*, p.233.
60 Ball, I.R. (1983) in *Syst. Zool.*, **32**:449.
61 Jardine, N. (1969) in *Biol. J. Linn. Soc.*, **1**:328.
62 Patterson, C. in *Darwin up to Date*, p.35.

63 Platnick, N.I. (1979) in *Syst. Zool.*, **28**:546.

64 Patterson echoes Platnick's claim in stating that '. . . the most important outcome of cladistics is that a simple, even naive method of discovering the groups of systematists – what used to be called the natural system – has led some of us to realize that much of today's explanation of nature, in terms of Neo-Darwinism, or the synthetic theory, may be empty rhetoric' [in *The Biologist*, **27**:239 (1980)].

65 See Webster, G.C. & Goodwin, B.C. (1982) 'The origin of species: a structuralist approach' in *J. Soc. Biol. Struct.*, **5**:15–47. They argue that evolutionary theory is deficient because it does not treat the organism as a real entity. They suggest, in its place, a 'Rational Morphology' which considers the individual organisms and the biological domain as a whole, to be *structures*, characterised in terms of *internal* relations. The structuralist element is concerned with order, its generation and transformation. In transformed cladistics, the emphasis on patterns is also indicative of structuralist sympathies. See also Brooks, D.R. (1981) 'Classifications as languages of empirical comparative biology' in *Advances in Cladistics. Proceedings of the first meeting of the Willi Hennig Society*, eds. V.A. Funk & D.R. Brooks. Bronx, New York: The New York Botanical Gardens.

66 Stanley, S.M. (1975) 'A theory of evolution above the species level' in *Proc. Nat. Acad. Sci.*, **72**:646–50; also (1979) *Macroevolution. Pattern and Process*. San Francisco: W.H. Freeman.

67 Dover, G.A. (1982) 'Molecular Drive: a cohesive mode of species evolution' in *Nature*, **299**:111–17; also Dover, G.A., Brown, S., Coen, E., Dallas, J., Strachan, T. & Trick, M. 'The dynamics of genome evolution and species differentiation' in *Genome Evolution*. Dover suggests that there is a third force in evolution, in addition to natural selection and genetic drift, namely molecular drive. This force describes DNA turnover (most noticeably in multigene families) in which a mutant member gene can gradually spread through a family in all individuals of a sexual population. The implication for evolutionary theory is the phenotype not only 'tracks' the environment (adaptation), but also *seeks* a new environment. [cf. Gould, S.J. & Vrba, E.S. (1982) 'Exaptation: a missing term in the science of form' in *Paleobiology*, **8**:4–15].

68 There is some evidence that this is beginning to occur. See Saether, O.A. (1983) 'The canalised evolutionary potential: inconsistencies in phylogenetic reasoning' in *Syst. Zool.*, **32**:343–59.

BIBLIOGRAPHY

Agassiz, L. (1857) 'Essay on classification' in *Contributions to the Natural History of the United States*, Vol. I. Boston: Little, Brown & Co.; Reprinted 1962, ed. C. Lurie. Cambridge, Mass.: Harvard Univ. Press.

Agassiz, L. (1859) *An Essay on Classification*. London: Longman, Brown, Green, Longmans, and Roberts and Trubner & Co.

Arnold, E.N. (1981) 'Estimating phylogenies at low taxonomic levels' in *Z.Zool. Syst. Evol.-forsch.*, 19:1–35.

Ashlock, P.D. (1974) 'The uses of cladistics' in *Ann. Rev. Ecol. & Syst.*, 5:81–99.

Ashlock, P.D. (1979) 'An evolutionary systematist's view of classification' in *Syst. Zool.*, 28:441–50.

Atran, S. (1983) 'Covert Fragmenta and the origins of the botanical family' in *Man*, 18:51–71.

Ball, I.R. (1981) 'The order of life – towards a comparative biology' in *Nature*, 294:675–6.

Ball, I.R. (1983) 'On groups, existence and the ordering of nature' in *Syst. Zool.*, 32:446–51.

Baroni-Urbani, C. (1977) 'Hologenesis, phylogenetic systematics, and evolution' in *Syst. Zool.*, 26:343–6.

Beatty, J. (1982) 'Classes and cladists' in *Syst. Zool.*, 31:25–34.

Beatty, J. (1983) 'What's in a word? Coming to terms in the Darwinian revolution' in *Nature Animated*, ed. M. Ruse. Dordrecht: D. Reidel

Beatty, J. & Fink, W.L. (1980) '[Review of] *Simplicity*' in *Syst. Zool.*, 28:643–51.

Beckner, M. (1959) *The Biological Way of Thought*. New York: Columbia Univ. Press.

Berlin, B., Breedlove, D.E. & Raven, P.H. (1966) 'Folk taxonomies and biological classification' in *Science*, 154:273–5.

Bernstein, R.J. (1983) *Beyond Objectivism and Relativism*. Oxford: Blackwell.

Bigelow, R.S. (1956) 'Monophyletic classification and evolution' in *Syst. Zool.*, 5:145–6.

Bigelow, R.S. (1958) 'Classification and phylogeny' in *Syst. Zool.*, 7:49–59.

Bigelow, R.S. (1961) 'Higher categories and phylogeny' in *Syst. Zool.*, 10:86–91.

Bishop, M.J. (1982) 'Criteria for the determination of the direction of character state changes' in *Zool. J. Linn. Soc.*, 74:197–206.

Bishop, M.J. & Friday, A.E. (1982) 'Evolutionary processes: two deterministic models, and their implications.' Unpublished manuscript.

Bock, W.J. (1974) 'Philosophical foundations of classical evolutionary classifications' in *Syst. Zool.*, **22**:375–92.

Bock, W.J. (1977) 'Foundations and methods of evolutionary classification' in *Major Patterns in Vertebrate Evolution*, eds. M.K. Hecht, P.C. Goody & B.M. Hecht. New York: Plenum Press.

Bonde, N. (1977) 'Cladistic classifications as applied to vertebrates ' in *Major Patterns in Vertebrate Evolution*, eds. M.K. Hecht, P.C. Goody & B.M. Hecht. New York: Plenum Press.

Boyd, R.N. (1973) 'Realism, underdetermination and a causal theory of evidence' in *Nous*, **7**:1–12.

Boyden, A. (1973) *Perspectives in Zoology*. Oxford: Pergamon Press.

Brady, R.H. (1982) 'Theoretical issues and pattern cladistics' in *Syst. Zool.*, **31**:286–91.

Brady, R.H. (1982) 'Parsimony, hierarchy and biological implications' in *Advances in Cladistics*, Vol. 2. *Proceedings of the second meeting of the Willi Hennig Society*, eds. N.I. Platnick & V.A. Funk. New York: Colombia Univ. Press.

Brandon, R. & Burian, R. (eds.) (1984) *Genes, Organisms, Populations*. Cambridge, Mass: MIT Press.

Bretsky, S.S. (1979) 'Recognition of ancestor–descendant relationships in vertebrate paleontology' in *Phylogenetic Analysis and Paleontology*, eds. J. Cracraft & N. Eldredge. New York: Columbia Univ. Press.

Bridgmann, W.P. (1927) *The Logic of Modern Physics*. New York: Macmillan.

Broadbent, D.E. (1961) *Behaviour*. London: Eyre & Spottiswoode.

Brooks, D.R. (1981) 'Classifications as languages of empirical comparative biology' in *Advances in Cladistics. Proceedings of the first meeting of the Willi Hennig Society*. Eds. V.A. Funk & D.R. Brooks. Bronx, New York: The New York Botanical Garden.

Brooks, D.R. (1982) 'Review of *Systematics and Biogeography*' in *Syst. Zool.*, **31**:206–8

Bruce, E.J. & Ayala, F.J. (1979) 'Phylogenetic relationships between man and the apes: electrophoretic evidence' in *Evolution*, **33**:1040–56.

Brundin, L. (1966) 'Transarctic relationships and their significance, as evidenced by chironomid midges' in *Kungla Svenska Vetensk. Akad. Handlingar*, **11**:1–472.

Brundin, L. (1968) 'Application of phylogenetic principles in systematics and evolutionary theory' in *Current Theories in Lower Vertebrate Phylogeny*. *Nobel Symposium 4*, ed. T. Ørvig. Stockholm: Almquist & Wiskell.

Brundin, L. (1972) 'Evolution, causal biology and classification' in *Zool. Scripta*, **1**:107–20.

Buck, R.C. & Hull, D.L. (1966) 'The logical structure of the Linnaean hierarchy' in *Syst. Zool.*, **15**:97–111.

Butler, P.M. (1982) 'Directions of evolution in the mammalian dentition' in *Problems of Phylogenetic Reconstruction. The Syst. Assoc. Special Vol. 21*, eds. K.A. Joysey & A.E. Friday. London: Academic Press.

Cain, A.J. (1954) *Animal Species and Evolution*. London: Hutchinson.

Cain, A.J. (1959) 'Deductive and inductive methods in post-Linnaean taxonomy' in *Proc. Linn. Soc. Lond.*, **170**:185–217.

Cain, A.J. (1959) 'Function and taxonomic importance' in *Function and Taxonomic Importance*, ed. A.J. Cain. London: Syst. Assoc. Publ. Vol. 2.

Cain, A.J. (1962) 'Zoological classification' in *Aslib. Proc.*, **14**:226–30.

Cain, A.J. & Harrison, G.A. (1958) 'An analysis of the taxonomist's judgement of affinity' in *Proc. Zool. Soc. Lond.*, **131**:85–98.

Cain, A.J. & Harrison, G.A. (1960) 'Phyletic weighting' in *Proc. Zool. Soc. Lond.*, **135**:1–31.

Caplan, A.L. (1980) 'Have species become declassé?' in *Proceedings of the Philosophy of Science Association 1980*, Vol. I, eds. P.A. Asquith & R.N. Giere. East Lansing, Michigan: Michigan Philosophy of Science Association.

Caplan, A.L. (1981) 'Back to class: a note on the ontology of species' in *Phil. Sci.*, 48:130–40.

Carson, H.L. (1970) 'Chromosome tracers of the origins of species: some Hawaiian *Drosophila* species have arisen from single founder individuals in less than a million years' in *Science*, 168:1414–8.

Cassidy, J. (1978) 'Philosophical aspects of the Group Selection Controversy' in *Phil. Sci.*, 45:575–94.

Chalmers, A.F. (1982) *What Is This Thing Called Science?* 2nd ed. Queensland: Univ. Queensland Press.

Charig, A.J. (1981) 'Cladistics: a different point of view' in *The Biologist*, 28:19–20.

Charig, A.J. (1982) 'Systematics in biology: a fundamental comparison of some major schools of thought' in *Problems of Phylogenetic Reconstruction. The Syst. Assoc. Special Vol. 21*, eds. K.A. Joysey & A.E. Friday. London: Academic Press.

Churchland, P.M. (1975) 'Two grades of evidential bias' in *Phil. Sci.*, 42:250–9.

Cracraft, J. (1974) 'Phylogenetic models and classification' in *Syst. Zool.*, 23:71–90.

Cracraft, J. (1978) 'Science, philosophy and systematics' in *Syst. Zool.*, 27:213–6.

Cracraft, J. (1981) 'Pattern and process in paleobiology: the role of cladistic analysis in systematic paleontology' in *Paleobiology*, 7:456–68.

Croziat, L. (1964) *Space, Time, Form: The Biological Synthesis.* Caracas: Published by the author.

Crowson, R.A. (1970) *Classification and Biology.* Atherton Press.

Darwin, C. (1959) *On the Origin of Species by natural selection, or the preservation of favoured races in the struggle for life.* London: John Murray.

Davidson, D. (1973–4) 'On the very idea of a conceptual scheme' in *Proc. & Adds. Am. Phil. Assoc.*, 47;5–20.

Dawkins, R. & Ridley, M. (eds) (1984) *Oxford Surveys in Evolutionary Biology.* Volume 1. Oxford University Press.

De Beer, G.R. (1928) *Vertebrate Zoology.* London: Sidgwick & Jackson.

Denton, M. (1985) *Evolution: A Theory in Crisis.* London: Burnett Books.

Dingle, H. (1938) 'The rational and empirical elements in physics' in *Philosophy*, 13:148–65.

Dobzhansky, Th. (1970) *The Genetics of the Evolutionary Process.* New York: Columbia Univ. Press.

Dobzhansky, Th., Ayala, F.J., Stebbins, G.L. & Valentine, J.W. (1977) *Evolution.* San Francisco: W.H. Freeman.

Dover, G.A. (1982) 'Molecular Drive: a cohesive mode of species evolution' in *Nature*, 299:111–17.

Dover, G.A., Brown, S., Coen, E., Dallas, J., Strachen, T. & Trick, M. (1982) 'The dynamics of genome evolution and species differentiation' in *Genome Evolution. The Syst. Assoc. Special Publ. Vol. 20*, eds. G.A. Dover & R.B. Flavell London: Academic Press.

Dover, G.A. & Flavell, R.B. (eds.) (1982) *Genome Evolution.* London: Academic Press.

Dummett, M. (1973) *Frege: Philosophy of Language.* London: Gerald Duckworth.

Dunbar, M.J. (1980) 'The blunting of Occam's Razor, or to hell with parsimony' in *Can. J. Zool.*, 58:123–8.

Duncan, T. & Baum, B.R. (1981) 'Numerical phenetics: its use in botanical systematics' in

Ann. Rev. Ecol. & Syst., **12**:387–404.

Duncan, T. & Stuessy, T.F. (eds.) (1984) Cladistics: Perspectives on the Reconstruction of Evolutionary History. New York: Columbia Univ. Press.

Dunn, G. & Everitt, B.S. (1982) An Introduction to Mathematical Taxonomy. Cambridge Univ. Press.

Dupraw, E.J. (1964) 'Non-Linnaean taxonomy' in Nature, **202**:849–52.

Dupraw, E.J. (1965) 'Non-Linnaean taxonomy and the systematics of honeybees' in Syst. Zool., **14**:1–24.

Dupraw, E.J. (1965) 'The recongition and handling of honeybee specimens in non-Linnaean taxonomy' in J. Apicult. Res., **4**:71–84.

Edwards, A. (1972) Likelihood. Cambridge: Cambridge Univ. Press.

Ehrlich, P.R. & Ehrlich, A.H. (1967) 'The phenetic relationships of butterflies' in Syst. Zool., **16**:301–17.

Eldredge, N (1985) Unfinished Synthesis: Biological Hierarchies and Modern Evolutionary Thought. New York: Oxford Univ. Press.

Eldredge, N. & Cracraft, J. (1980) Phylogenetic Patterns and the Evolutionary Process. New York: Columbia Univ. Press.

Eldredge, N. & Gould, S.J. (1972) 'Punctuated equilibria: an alternative to phyletic gradualism' in Models in Paleobiology, ed. T.J.M. Schopf. San Francisco: Freeman Cooper.

Eldredge, N & Tattershall I. (1977) 'Evolutionary models, phylogenetic reconstruction, and another look at Hominid phylogeny' in Historical Biogeography, Plate Tectonics and the Changing Environment, eds. J. Gray & A.J. Boucot. Corvallis: Oregon State Univ. Press, pp. 147–67.

Engelmann, G.F. & Wiley, E.O. (1977) 'The place of ancestor–descendent relationships in phylogeny' in Syst. Zool., **26**:1–11.

Estabrook, G.F. (1972) 'Cladistic methodology: a discussion of the theoretical basis for the induction of evolutionary history' in Ann. Rev. Ecol. & Syst., **3**:427–56.

Estabrook, G.F., Johnson, C.S. & McMorris, F.R. (1975) 'An idealised concept of the true cladistic character' in Math. Biosc., **23**:263–72.

Farris, J.S. (1970) 'Methods for computing Wagner trees' in Syst. Zool., **19**:83–92.

Farris, J.S. (1977) 'On the phenetic approach to vertebrate classification' in Major Patterns in Vertebrate Evolution, eds. M.K. Hecht, P.C. Goody & B.M. Hecht. New York: Plenum Press.

Farris, J.S. (1977) 'Phylogenetic analysis under Dollo's law' in Syst. Zool., **26**:77–88.

Farris, J.S. (1979) 'The information content of the phylogenetic system' in Syst. Zool., **28**:483–519.

Farris, J.S. (1982) 'Simplicity and informativeness in systematics and phylogeny' in Syst. Zool., **31**:413–44.

Farris, J.S. (1983) 'The logical basis of phylogenetic analysis' in Advances in Cladistics, Vol. II. Proceedings of the second meeting of the Willi Hennig Society, eds. N.E. Platnick & V.A. Funk. New York: Columbia Univ. Press.

Farris, J.S. (1985) 'Another kind of Cladistics' in Cladistics, **1**:292–9.

Felsenstein, J. (1978) 'The number of evolutionary trees' in Syst. Zool., **27**:27–33.

Felsenstein, J. (1978) 'Cases in which parsimony or compatibility will be positively misleading' in Syst. Zool., **27**:401–10.

Felsenstein, J. (1982) 'Numerical methods for inferring evolutionary trees' in *Quart. Rev. Biol.*, **57**:379–404.

Felsenstein, J. (1983) 'Parsimony in systematics: biological and statistical issues' in *Ann. Rev. Ecol. & Syst.*, **14**:313–33.

Feyerabend, P.K. (1962) 'Explanation, reduction and empiricism' in *Minnesota Studies in the Philosophy of Science*, Vol. III, eds. H. Feigl & G. Maxwell. Minneapolis: Univ. Minnesota Press, pp. 28–97.

Feyerabend, P.K. (1975) *Against Method*. London: Verso.

Feyerabend, P.K. (1981) *Realism, Rationalism and Scientific Method. Philosophical Papers*, Vol. I. Cambridge: Cambridge Univ. Press.

Feyerabend, P.K. (1981) *Problems of Empiricism. Philosophical Papers*, Vol. II. Cambridge: Cambridge Univ. Press.

Fink, S.V. (1982) 'Report on the second annual general meeting of the Willi Hennig Society' in *Syst. Zool.*, **31**:180–97.

Forey, P.L. (1982) 'Neontological analysis versus paleontological stories' in *Problems of Phylogenetic Reconstruction. The Syst. Assoc. Special Vol. 21*, eds. K.A. Joysey & A.E. Friday. London: Academic Press.

Forster, M.R. (1986) 'Statistical Covariance as a Measure of Phylogenetic Relationship' in *Cladistics*, **2** (4):297–317.

Friday, A.E. (1982) 'Parsimony, simplicity and what actually happened' in *Zool. J. Linn. Soc.*, **74**:329–35.

Gaffney, E.S. (1979) 'An introduction to the logic of phylogeny reconstruction' in *Phylogenetic Analysis and Paleontology*, eds. J. Cracraft & N. Eldredge. New York: Columbia Univ. Press.

Ghiselin, M.T. (1966) 'On psychologism in the logic of taxonomic principles' in *Syst. Zool.*, **15**:207–15.

Ghiselin, M.T. (1969) *The Triumph of the Darwinian Method*. Berkeley: Univ. California Press.

Ghiselin, M.T. (1974) 'A radical solution to the species problem' in *Syst. Zool.*, **23**:536–44.

Ghiselin, M.T. (1984) ' 'Definition', 'character', and other equivocal terms' in *Syst. Zool.*, **33**:104–10.

Gilmour, J.S.L. (1937) 'A taxonomic problem' in *Nature*, **139**:1040–2.

Gilmour, J.S.L. (1940) 'Taxonomy and philosophy' in *The New Systematics*, ed. J.S. Huxley. Oxford: Oxford Univ. Press.

Gilmour, J.S.L. (1961) 'Taxonomy' in *Contemporary Botanical Thought*, eds. A.M. MaCleod & C.S. Cobley. Edinburgh: Oliver & Boyd.

Gilmour, J.S.L. & Walters, S.M. (1963) 'Philosophy and classification' in *Vistas in Botany*, IV:1–22, ed. W.B. Turrill. London: Pergamon Press.

Gingerich, P.D. (1979) 'The stratophenetic approach to phylogeny reconstruction in vertebrate paleontology' in *Phylogenetic Analysis and Paleontology*, eds. J. Cracraft & N. Eldredge. New York: Columbia Univ. Press.

Gish, D.T. (1978) *Evolution – The Fossils Say No!* San Diego: Creation Life Publishers.

Goldschmidt, R.B. (1952) 'Evolution as viewed by one geneticist' in *Amer. Sci.*, **40**:84–98.

Goodman, M. & Moore, G.W. (1971) 'Immunodiffusion systematics of the primates I. The Catarrhini' in *Syst. Zool.*, **20**:19–62.

Goodman, N. (1954) *Fact, Fiction and Forecast*. London: Athlone Press.

Gorman, G.C., Wilson, A.C. & Nakanishi, M. (1971) 'A biochemical approach towards the study of reptilian phylogeny. Evolution of serum albumin and lactic dehydrogenase in *Syst. Zool.*, 20:167–85.

Gould, S.J. (1980) *Ontology and Phylogeny.* Cambridge, Mass.: Harvard Univ. Press.

Gould, S.J. (1980) 'G.G. Simpson, paleontology and the modern synthesis' in *The Evolutionary Synthesis*, eds. E. Mayr & W.B. Provine. Cambridge, Mass.: Harvard Univ. Press.

Gould, S.J. (1980) 'The promise of paleobiology as a nomothetic, evolutionary discipline' in *Paleobiology*, 6:96–118.

Gould, S.J. (1980) 'Is a new and general theory of evolution emerging?' in *Paleobiology*, 6:119–30.

Gould, S.J. (1982) 'Punctuated equilibria – a new way of seeing' in *New Scientist*, 94:137–41.

Gould, S.J. & Eldredge, N. (1977) 'Punctuated equilibria: the tempo and mode of evolution reconsidered' in *Paleobiology*, 3:115–51.

Gould, S.J. & Lewontin, R. (1979) 'The spandrels of San Marco and the Panglossian paradigm: a critique of the adaptationist programme' in *Proc. Roy. Soc. Ser. B*, 205:581–98.

Gould, S.J. & Vrba, E.S. (1982) 'Exaptation: a missing term in the science of form' in *Paleobiology*, 8:4–15.

Gregg, J.R. (1954) 'Taxonomy, language and reality' in *Am. Nat.*, 84:419–35.

Gregg, J.R. (1954) *The Language of Taxonomy.* New York: Columbia Univ. Press.

Gregg, J.R. (1967) 'Finite Linnaean structures' in *Bull. Math. Biophys.*, 29:191–266.

Grene, M. (1983) (ed.) *Dimensions of Darwinism.* Cambridge: Cambridge Univ. Press.

Griffiths, G.C.D. (1974) 'On the foundation of biological systematics' in *Acta Biotheoretica*, 23:85–131.

Griffiths, G.C.D. (1976) 'The future of Linnaean nomenclature' in *Syst. Zool.*, 25:168–73.

Guédès, M. (1967) 'La méthode taxonomique d'Adanson' in *Rev. Hist. Sci.*, 20:361–86.

Haas, O. & Simpson, G.G. (1946) 'Analysis of some phylogenetic terms with attempts at redefinition' in *Proc. Am. Phil. Sci.*, 90:319–49.

Hacking, I. (1983) *Representing and Intervening.* Cambridge: Cambridge Univ. Press.

Halstead, L.B. (1982) 'Evolutionary trends and the phylogeny of the agnatha' in *Problems of Phylogenetic Reconstruction. The Syst. Assoc. Special Vol. 21*, eds. K.A. Joysey & A.E. Friday. London: Academic Press.

Hare, R.M. (1982) *Plato.* Oxford: Oxford Univ. Press.

Hecht, M.K. & Edwards, J.L. (1977) 'The methodology of phylogenetic inference above the species level' in *Major Patterns in Vertebrate Evolution*, eds. M.K. Hecht, P.C. Goody, & B.M. Hecht. New York: Plenum Press.

Hempel, C.G. (1965) *Aspects of Scientific Explanation.* New York: MacMillan.

Hennig, W. (1965) 'Phylogenetic systematics' in *Ann. Rev. Entomol.*, 10:97–116.

Hennig, W. (1966) *Phylogenetic Systematics.* Urbana: Univ. Illinois Press.

Hennig, W. (1975) 'Cladistics analysis or cladistic classification? A reply to Ernst Mayr' in *Syst. Zool.*, 24:244–56.

Hesse, M.B. (1974) *The Structure of Scientific Inference.* Macmillan.

Hesse, M.B. (1980) *Revolutions and Reconstructions in the Philosophy of Science.* Brighton: Harvester Press.

Hill, C.R. & Crane, P.R. (1982) 'Evolutionary cladistics and the origins of angiosperms' in *Problems of Phylogenetic Reconstruction. The Syst. Assoc. Special Vol. 21*, eds. K. A. Joysey & A.E. Friday. London: Academic Press.

Hull, D.L. (1965) 'The effect of essentialism on taxonomy - Two thousand years of stasis' in *Brit. J. Phil. Sci.*, 15:314-26; 16:1-18.

Hull, D.L. (1967) 'Certainty and circularity in evolutionary taxonomy' in *Evolution*, 21:174-89.

Hull, D.L. (1968) 'The operational imperative: sense and nonsense in operationalism' in *Syst. Zool.*, 16:438-57.

Hull, D.L. (1970) 'Contemporary systematic philosophies' in *Ann. Rev. Ecol. & Syst.*, 1:19-54.

Hull, D.L. (1976) 'Are species really individuals?' in *Syst. Zool.*, 25:174-91.

Hull, D.L. (1978) 'A matter of individuality' in *Phil. Sci.*,45:335-60.

Hull, D.L. (1979) 'The limits of cladism' in *Syst. Zool.*, 28:416-40.

Hull, D.L. (1981) 'The principles of biological classification: the use and abuse of philosophy' in *Proceedings of the Philosophy of Science Association 1978*, Vol. II, eds. P.A. Asquith & I. Hacking. East Lansing, Michigan: Michigan Philosophy of Science Association.

Hull, D.L. (1981) 'Kitts and Kitts and Caplan on species ' in *Phil. Sci.*, 48:141-52.

Hull, D.L. (1983) 'Darwin and the nature of science' in *Evolution from Molecules to Man*, ed. D.S. Bendall. Cambridge Univ. Press.

Hull, D.L. (1984) 'Can Kripke alone save Essentialism? A Reply to Kitts' in *Syst. Zool.*, 33:110-12.

Humphries, C.J. (1983) 'Review of *Problems of Phylogenetic Reconstruction* and *Methods of Phylogenetic Reconstruction*' in *Syst. Zool.*, 32:302-10.

Huxley, J.S. (1940) (ed) *The New Systematics*. Oxford: Clarendon Press.

Huxley, J.S. (1958) 'Evolutionary processes and taxonomy with special reference to grades' in *Uppsala. Univ. Arssk.*, 6:21-39.

Huxley, J.S. (1959) 'Clades and grades' in *Function and Taxonomic Importance*, ed. A.J. Cain. Syst. Assoc. Publ., Vol. 3.

Inglis, W.G. (1966) 'The observational basis of homology' in *Syst. Zool.*, 15:219-28.

Ions, E. (1977) *Against Behaviourism. A Critique of Behavioural Science*. Oxford: Basil Blackwell.

Janis, C. (1982) 'Evolution of horns in Ungulates: ecology and paleoecology' in *Biol. Rev.*, 57:261-318.

Jardine, N. (1967) 'The concept of homology in biology' in *Brit. J. Phil. Sci.*, 18:125-39.

Jardine, N. (1969) 'The observational and theoretical components of homology: a study based on the dermal roofs of rhipidistian fishes' in *Biol. J. Linn. Soc.*, 1:327-61.

Jardine, N. (1969) Studies in the Theory of Classification. Ph.D. thesis. Cambridge University.

Jardine, N. (1969) 'A logical basis for biological classification' in *Syst. Zool.*, 18:37-52.

Jardine, N. (1971) 'Patterns of differentiation between human local populations' in *Phil. Trans. Roy. Soc. Lond.*, B, 263:1-33.

Jardine, N. & Sibson, R. (1968) 'The construction of hierarchic and non-hierarchic classifications' in *Computer Journal*, 11:177-84.

Jardine, N. & Sibson, R. (1971) *Mathematical Taxonomy*. London: John Wiley.

Jardine, N., Van Rijsbergen, C.J. & Jardine, C.J. (1969) 'Evolutionary rates and the inference of evolutionary tree forms' in *Nature*, 244:185.

Jarvik, E. (1948) 'On the morphology and taxonomy of the Middle Devonian osteolepid fishes of Scotland' in *Kungla Svenska. Vetensk. Akad. Handlingar*, 25:1-301.

Jefferies, R.P.S. (1979) 'The origin of chordates - a methodological essay' in *The Origin of*

Major Invertebrate Groups, ed. M.R. House. London: Academic Press.

Johnson, L.A.S. (1968) 'Rainbow's end: the quest for an optimal taxonomy' in *Proc. Linn. Soc. N.S.W.*, **93**:8–45.

King, M.C. & Wilson, A.C. (1975) 'Evolution at two levels in humans and chimpanzees' in *Science*, **188**:107–16.

Kiriakoff, S. (1959) 'Phylogenetic systematics versus typology' in *Syst. Zool.*, **8**:117–18.

Kitcher, P. (1982) *Abusing Science. The Case against Creationism.* Cambridge, Mass.: The MIT Press.

Kitcher, P. (1984) 'Species' in *Phil. Sci.*, **51**:308–33.

Kitts, D.B. (1977) 'Karl Popper, verifiability and systematic zoology' in *Syst. Zool.*, **26**:185–94.

Kitts, D.B. (1978) 'Theoretics and systematics: a reply to Cracraft, Nelson and Patterson', in *Syst. Zool.*, **27**:222–4.

Kitts, D.B. (1984) 'The Names of Species: A Reply to Hull' in *Syst. Zool.*, **33**:112–15.

Kitts, D.B. & Kitts, D.J. (1979) 'Biological species as natural kinds' in *Phil. Sci.*, **46**:613–22.

Kohne, D.E., Chicson, J.A. & Hoyer, B.H. (1972) 'Evolution of primate DNA sequences' in *J. Human Evol.*, **1**:627–44.

Kripke, S. (1972) 'Naming and necessity' in *Semantics of Natural Languages*, eds. G. Harman & D. Davidson. Dordrecht: D. Reidel.

Kuhn, T.S. (1962) *The Structure of Scientific Revolutions. International Encyclopedia of Unified Science*, Vol. 2, No. 2. Chicago: Univ. Chicago Press.

Lakatos, I & Musgrave, A. (eds.) (1970) *Criticism and the Growth of Knowledge.* Cambridge: Cambridge Univ. Press.

Lance, G.N. & Williams, W.T. (1965) 'Computer programs for monothetic classification (association analysis)' in *Computer Journal*, **8**:246–9.

Laudan, L. (1977) *Progress and its Problems.* London: Routledge & Kegan Paul.

Lauder, G.V. (1981) 'Form and Function. Structural analysis in evolutionary morphology' in *Paleobiol.*, **7**:430–42.

Leith, B. (1982) *The Descent of Darwin.* London: Collins.

LeQuesne, W.J. (1969) 'A method for selecting characters in numerical taxonomy' in *Syst. Zool.*, **18**:201–5.

Lewin, R. (1980) 'Evolutionary theory under fire' in *Science*, **210**:883–7.

Lewontin, R.C. (1974) *The Genetic Basis of Evolutionary Change.* New York: Columbia Univ. Press.

Little, J. (1980) 'Evolution: myth, metaphysics or science?' in *New Scientist*, **87**:708–9.

Lockhart, W.R. & Hartmann, P.A. (1963) 'Formation of monothetic groups in quantitative bacterial taxonomy' in *J. Bacteriol.*, **85**:68–77.

Løvtrup, S. (1973) 'Classification, convention and logic' in *Zool. Scripta.*, **2**:49–61.

Løvtrup, S. (1974) *Epigenetics, a Treatise on Theoretical Biology.* New York: John Wiley.

Løvtrup, S. (1977) *The Phylogeny of Vertebrata.* New York: John Wiley.

Løvtrup, S. (1981) 'Macroevolution and punctuated equilibria' in *Syst. Zool.*, **30**:498–500.

Macgregor, H.C. (1982) 'Big chromosomes and speciation amongst amphibia' in *Genome Evolution. Syst. Assoc. Spec. Publ. Vol. 20*, eds. G.A. Dover & R.B. Flavell. London: Academic Press.

Mayr, E. (1942) *Systematics and the Origins of Species.* New York: Columbia Univ. Press.

Mayr, E. (1963) *Animal Species and Evolution.* Cambridge, Mass.: Harvard Univ. Press.

Mayr, E. (1965) 'Numerical phenetics and taxonomic theory' in *Syst. Zool.*, **14**:73–97.

Mayr, E. (1968) 'The theory of biological classification' in *Nature*, 220:545-8.

Mayr, E. (1969) *Principles of Systematic Zoology*. New York: McGraw-Hill.

Mayr, E. (1974) 'Cladistic analysis or cladistic classification?' in *Z. Zool. Syst. Evol.-forsch*, 12:94-128.

Mayr, E. (1975) 'The unity of the genotype' in *Biologische Zentralblatt*, 94:377-88.

Mayr, E. (1976) *Evolution and the Diversity of Life: Selected Essays*. Cambridge, Mass.: Harvard Univ. Press.

Mayr, E. (1981) 'Biological classification: toward a synthesis of opposing methodologies' in *Science*, 214:510-6.

Mayr, E. (1982) *The Growth of Biological Thought*. Cambridge, Mass.: Harvard Univ. Press.

Mayr, E. (1984) 'Typological versus population thinking' in *Conceptual Issues in Evolutionary Biology*, ed. E. Sober. Cambridge, Mass.: MIT Press.

Mayr, E. Linsley, E.G. & Usinger, R.L. (1953) *Methods and Principles of Systematic Zoology*. New York: McGraw-Hill.

McKenna, M.C. (1975) 'Toward a phylogenetic classification of the Mammalia' in *Phylogeny of the Primates*, eds. W.P. Luckett & F.S. Szalay. New York: Plenum Press.

McNeill, J. (1979) 'Phylogenetic reconstruction and phenetic taxonomy' in *Zool. J. Linn. Soc.*, 74:337-44.

McNeill, J. (1982) 'Purposeful phenetics' in *Syst. Zool.*, 28:465-82.

Michener, C.D. (1963) 'Some future developments in taxonomy' in *Syst. Zool.*, 12:151-72.

Michener, C.D. (1970) 'Diverse approaches to systematics' in *Evolutionary Biology*, 4, eds. Th. Dobzhansky, M.K. Hecht & M.C. Steere. New York: Appleton-Century-Crofts.

Mickevich, M.C. (1978) 'Taxonomic congruence' in *Syst. Zool.*, 27:143-58.

Mill, J.S. (1974) 'A synthesis of logic ratiocinative and inductive. Books I-VII:IV: Of operations subsidiary to induction' in *Collected Works of John Stuart Mill*, Vol. IV, ed. J. Robson. Toronto: Univ. Toronto Press.

Mitchell, P.C. (1901) 'On the intestinal tract of birds: with remarks on the valuation and nomenclature of zoological characters' in *Trans. Linn. Soc. Lond.*, Series 2, 8:173-275.

Moss, W.W. (1972) 'Some levels of phenetics' in *Syst. Zool.*, 21:236-9.

Nagel, E. (1961) *The Structure of Science*. London: Routledge & Kegan Paul.

Nelson, G.J. (1970) 'Outline of a theory of comparative biology' in *Syst. Zool.*, 19:373-84.

Nelson, G.J. (1971) 'Cladism as a philosophy of science' in *Syst. Zool.*,20:373-6.

Nelson, G.J. (1972) 'Phylogenetic relationship and classification' in *Syst. Zool.*, 21:227-30.

Nelson, G.J. (1972) 'Comments on Hennig's *Phylogenetic Systematics* and its influence on ichthyology' in *Syst. Zool.*, 21:364-74.

Nelson, G.J. (1973) 'The higher level phylogeny of vertebrates' in *Syst. Zool.*, 22:87-91.

Nelson, G.J. (1973) 'Monophyly again? - a reply to P.D. Ashlock' in *Syst. Zool.*, 22:310-12.

Nelson, G.J. (1973) 'Classification as an expression of phylogenetic relationships' in *Syst. Zool.*, 22:344-59,

Nelson, G.J. (1978) 'Classification and prediction - a reply to Kitts' in *Syst. Zool.*, 27:216-18.

Nelson, G.J. (1978) 'Ontology, phylogeny and the biogenetic law' in *Syst. Zool.*, 27:324-45.

Nelson, G.J. (1979) 'Cladistic analysis and synthesis: principles and definitions with a historical note on Adanson's *Familles des Plantes* (1763-1764)' in *Syst. Zool.*, 28:1-21.

Nelson, G.J. & Platnick, N.I. (1981) *Systematics and Biogeography*. New York: Columbia Univ. Press.

Nelson, G.J. & Platnick, N.I. (1984) 'Systematics and Evolution' in *Beyond Neo-Darwinism*, eds. M.-W. Ho & P.T. Saunders. London: Academic Press, pp. 143–58.

Newell, N.B. (1959) 'The nature of the fossil record' in *Proc. Am. Phil. Sci.*, **103**:264–85.

Newton-Smith, W.H. (1978) 'The underdetermination of theory by data' in *Aristotelean Society* Supplementary Volume LII, pp. 71–91.

Newton-Smith, W.H. (1981) *The Rationality of Science*. London: Routledge & Kegan Paul.

Ospovat, D. (1981) *The Development of Darwin's Theory*. Cambridge Univ. Press.

Owen, R. (1849) *On the Nature of Limbs*. London: John Van Voorst.

Panchen, A.L. (1982) 'The use of parsimony in testing phylogenetic hypotheses' in *Zool. J. Linn. Soc.*, **74**:305–28.

Paterson, H.E.H. (1978) 'More Evidence against speciation by reinforcement' in *S. Af. J. Sci.*, **77**:369–71.

Paterson, H.E.H. (1982) 'Perspectives on speciation by reinforcement' in *S. Af. J. Sci.*, **78**:53–7.

Paterson, H.E.H. (1985) 'The recognition concept of species' in *Species and Speciation*, ed. E.S. Vrba. *Transvaal Museum Monograph No. 4*, Transvaal Museum, Pretoria, pp. 21–29.

Patterson, C. (1977) 'The contribution of paleontology to teleostean phylogeny' in *Major Patterns in Vertebrate Evolution*, eds. M.K. Hecht, P.C. Goody & B.M. Hecht. New York: Plenum Press.

Patterson, C. (1978) 'Arthropods and ancestors' in *Antenna*, **2**:93–103.

Patterson, C. (1978) 'Verifiability in systematics' in *Syst. Zool.*, **27**:218–22.

Patterson, C. (1980) 'Cladistics' in *The Biologist*, **27**:234–40.

Patterson, C. (1980) 'Origins of tetrapods: historical introduction to the problem' in *The Terrestrial Environment and the Origin of Land Vertebrates*. Syst. Assoc. Special Vol. 15, ed. A.L. Panchen. London: Acadademic Press.

Patterson, C. (1981) 'Significance of fossils in determining evolutionary relationships' in *Ann. Rev. Ecol. & Syst.*, **12**:195–223.

Patterson, C. (1982) 'Cladistics and classification' in *Darwin up to Date*, ed. J. Cherfas. IPC Magazines.

Patterson, C. (1982) 'Classes and cladists or individuals and evolution' in *Syst. Zool.*, **31**:284–6.

Patterson, C. (1982) 'Morphological characters and homology' in *Problems of Phylogenetic Reconstruction*. Syst. Assoc. Special Vol. 21, eds. K.A. Joysey & A.E. Friday. London: Academic Press.

Patterson, C. (1983) 'How does phylogeny differ from ontology?' in *Development and Evolution*, eds. B.C. Goodwin, N. Holder & C.C. Wylie. Cambridge Univ. Press.

Patterson, C. & Rosen, D.E. (1977) 'Review of the ichthyodectiform and other Mesozoic teleost fishes and the theory and practice of classifying fossils' in *Bull. Am. Mus. Nat. Hist.*, **158**:81–172.

Paul, C.R.C. (1982) 'The adequacy of the fossil record' in *Problems of Phylogenetic Reconstruction*. Syst. Assoc. Special Vol. 21, eds. K.A. Joysey & A.E. Friday. London: Academic Press.

Platnick, N.I. (1977) 'Review of concepts of species' in *Syst. Zool.*, **26**:96–8.

Platnick, N.I. (1977) 'Cladograms, phylogenetic trees, and hypothesis testing' in *Syst. Zool.*, **26**:438–42.

Platnick, N.I. (1978) 'Gaps and prediction in classification' in *Syst. Zool.*, **27**:472–4.
Platnick, N.I. (1979) 'Philosophy and the transformation of cladistics' in *Syst. Zool.*, **28**:537–46.
Platnick, N.I. (1982) 'Defining characters and evolutionary groups' in *Syst. Zool.*, **31**:282–4.
Platnick, N.I. (1986) 'Evolutionary cladistics or evolutionary systematics?' in *Cladistics* **2**:288–96.
Platnick, N.I. & Cameron, H. (1977) 'Cladistic methods in textual, linguistic, and phylogenetic analysis' in *Syst. Zool.*, **26**:380–5.
Platnick, N.I. & Funk, V.A. (1982) *Advances in Cladistics Vol. II. Proceedings of the second meeting of the Willi Hennig Society*. New York: Columbia Univ. Press.
Platnick, N.I. & Gaffney, E.S. (1977) 'Systematics: a Popperian perspective' in *Syst. Zool.*, **26**:360–5.
Platnick, N.I. & Gaffney, E.S. (1978) 'Evolutionary biology: a Popperian perspective' in *Syst. Zool.*, **27**:131–41.
Platnick, N.I. & Gaffney, E.S. (1978) 'Systematics and the Popperian paradigm' in *Syst. Zool.*, **27**:381–8.
Platnick, N.I. & Nelson, G.J. (1978) 'A method of analysis for historical biogeography' in *Syst. Zool.*, **27**:1–17.
Platnick, N.I. & Nelson, G.J. (1981) 'The purpose of biological classification' in *Proceedings of the Philosophy of Science Association 1978*, Vol. II, eds. P.D. Asquith & I. Hacking. East Lansing, Michigan: Michigan Philosophy of Science Association.
Platts, M. (1983) 'Explanatory kinds' in *Brit. J. Phil. Sci.*, **34**:133–48.
Popper, K.R. (1959) *The Logic of Scientific Discovery*. London: Hutchinson.
Popper, K.R. (1959) *The Poverty of Historicism*. London: Routledge & Kegan Paul.
Popper, K.R. (1972) *Objective Knowledge. An Evolutionary Approach*. Oxford Univ. Press.
Popper, K.R. (1974) 'Darwinism as a metaphysical research programme' in *The Philosophy of Karl Popper*, ed. P.A. Schlipp. La Salle, Illinois: Open Court I.
Putnam, H. (1962) 'What theories are not' in *Logic, Methodology and Philosophy of Science*, ed. E. Nagel. Stanford, California: California Univ. Pres.
Putnam, H. (1975) *Mind, Language and Reality. Philosophical Papers, Vol. 2*. Cambridge Univ. Press.
Putnam, H. (1982) 'Why Reason Can't be Naturalised' in *Synthese*, **52**:3–23.
Quine, W.V. (1969) *Ontological Relativity and Other Essays*. New York: Columbia University Press.
Ridley, M. (1983) 'Can classification do without evolution?' in *New Scientist*, **100**:647–51.
Ridley, M. *Evolution and Classification*. London: Longman.
Rohlf, F.J. & Sokal, R.R. (1980) 'Comments on taxonomic congruence' in *Syst. Zool.*, **29**:97–101.
Romer, A.S. & Parsons, T.S. (1977) *The Vertebrate Body*. Philadelphia: W.B. Saunders, 5th edn.
Romero-Herrera, A.E., Lehman, H., Joysey, K.A. & Friday, A.E. (1973) 'Molecular evolution of myoglobin and the fossil record: a phylogenetic synthesis' in *Nature*, **246**:389–95.
Romero-Herrera, A.E., Lehman, H., Joysey, K.A. & Friday, A.E. (1978) 'On the evolution of myoglobin' in *Phil. Trans. Roy. Soc. Lond.*, B, **283**:61–163.
Rorty, R.J. (1980) *Philosophy and the Mirror of Science*. Princetown: Princetown Univ. Press.

Rosen, D.E. (1982) 'Do current theories of evolution satisfy the basic requirements of explanation?' in *Syst. Zool.*, **31**:76–85.

Rosen, D.E., Forey, P.L., Gardiner, B.G. & Patterson, C. (1981) 'Lungfishes, tetrapods, paleontology and plesiomorphy' in *Bull. Am. Mus. Nat. Hist.*, **167** (4):159–276.

Ross, H.H. (1974) *Biological Systematics*. Reading, Mass.: Addison-Wesley.

Ruse, M. (1973) *The Philosophy of Biology*. London: Hutchinson & Co.

Ruse, M. (1977) 'Karl Popper's philosophy of biology' in *Phil. Sci.*, **44**:638–61.

Ruse, M. (1979) 'Falsifiability, consilience and systematics' in *Syst. Zool.*, **28**:530–6.

Ruse, M. (1982) *Darwinism Defended. A Guide to the Evolution Controversies*. London: Addison-Wesley.

Saether, O.A. (1983) 'The cannalised evolutionary potential: inconsistencies in phylogenetic reasoning' in *Syst. Zool.*, **32**:343–59.

Sattler, R. (1964) 'Methodological problems in taxonomy' in *Syst. Zool.*, **13**:19–27.

Sattler, R. (1966) 'Towards a more adequate approach to comparative morphology' in *Phytomorphology*, **16**:417–29.

Sattler, R. (1967) 'Petal inception and the problem of pattern detection' in *J. Theor. Biol.*, **17**:31–9.

Säve-Söderbergh, G. (1934) 'Some points of view concerning the evolution of vertebrates and the classification of this group' in *Ark. Zool.*, **26A**:1–20.

Schaeffer, B., Hecht, M.K. & Eldredge, N. (1972) 'Phylogeny and paleontology' in *Evolutionary Biology*, 6, eds. T. Dobzhansky, M.K. Hecht & W.C. Steere. New York: Appleton-Century-Crofts.

Scheffler, I. (1967) *Science and Subjectivity*. Indianapolis: Bobbs-Merrill.

Schlee, D. (1971) 'Die rekonstruktion der phylogenese mit Hennig's prinzip' in *Aufsätze u. Reden. Senckenberg. Naturforsch. Ges.*, **20**:1–62.

Schnell, G.D. (1970) 'A phenetic study of the suborder Lari (Aves)' in *Syst. Zool.*, **19**:264–302.

Scott-Ram, N.R. (1982) Neo-Darwinism under attack: Is there any evidence of a revolution in the contemporary evolutionary debate? M. Phil. thesis. Cambridge University.

Sellars, W. (1963) *Science, Perception and Reality*. London: Routledge & Kegan Paul.

Settle, T. (1979) 'Popper on when is a science not a science?' in *Syst. Zool.*, **28**:521–9.

Simpson, G.G. (1944) *Tempo and Mode in Evolution*. New York: Columbia Univ. Press.

Simpson, G.G. (1951) 'The species concept in evolution' in *Evolution*, **5**:285–98.

Simpson, G.G. (1961) *Principles of Animal Taxonomy*. New York: Columbia Univ. Press.

Simpson, G.G. (1963) 'The meaning of taxonomic statements' in *Classification and Human Evolution*, ed. S.L. Washburn. Chicago: Aldine.

Simpson, G.G. (1963) 'Historical science' in *The Fabric of Geology*, ed. C.L. Albritton. Stanford, California: Freeman, Cooper & Co.

Simpson, G.G. (1976) 'The compleat paleontologist' in *Ann. Rev. Earth Planet Sci.*, **4**:1–13.

Simpson, G.G. (1978) 'Variation and details of macroevolution' in *Paleobiology*, **4**:217–21.

Skinner, B.F. (1950) 'Are theories of learning necessary?' in *Psychol. Rev.*, **57**:193–216.

Skinner, B.F. (1957) *Verbal Behaviour*. New York: Appleton-Century-Crofts.

Sneath, P.H.A. (1961) 'Recent developments in theoretical and quantitative taxonomy' in *Syst. Zool.*, **10**:118–39.

Sneath, P.H.A. (1964) 'Mathematics and classification from Adanson to the present' in *Adanson. The Bicentennial of Michel Adanson's "Familles des Plantes"*, ed. G.H.M. Lawrence. Pittsburg: Carnegie Institute of Technology, part 2, pp. 471–98.

Sneath, P.H.A. (1967) 'Trend-surface analysis of transformation grids' in *J. Zool., Lond.*, 151:65–122.

Sneath, P.H.A. & Sokal, R.R. (1973) *Numerical Taxonomy*. San Francisco: W.H. Freeman.

Sober, E. (1975) *Simplicity*. Oxford: Clarendon Press.

Sober, E. (1980) 'Evolution, population thinking and essentialism' in *Phil. Sci.*, 47:350–83.

Sober, E. (1982) 'The principles of parsimony' in *Brit. J. Phil. Sci.*, 32:145–56.

Sober, E. (1981) 'Evolutionary theory and the ontological status of properties' in *Phil. Studies*, 40:147–76.

Sober, E. (1983) 'Parsimony in systematics: philosophical issues' in *Ann. Rev. Ecol. & Syst.*, 14:335–57.

Sober, E. (ed.) (1984) *Conceptual Issues in Evolutionary Biology*. Cambridge, Mass.: MIT Press.

Sober, E. (1985) 'A likelihood justification of parsimony' in *Cladistics*, 1:209–33.

Sokal, R.R. (1962) 'Typology and empiricism in taxonomy' in *J. Theor. Biol.*, 3:230–67.

Sokal, R.R. (1966) 'Numerical taxonomy' in *Scient. Am.*, 215:106–16.

Sokal, R.R. & Camin, J.H. (1965) 'The two taxonomies: areas of agreement and conflict' in *Syst. Zool.*, 14:176–95.

Sokal, R.R. & Rohlf, F.J. (1970) 'The intelligent ignoramus: an experiment in numerical taxonomy' in *Taxon*, 19:305–19.

Sokal, R.R. & Sneath, P.H.A. (1963) *The Principles of Numerical Taxonomy*. San Francisco: W.H. Freeman.

Stanley, S.M. (1975) 'A theory of evolution above the species level' in *Proc. Nat. Acad. Sci.*, 72: 646–50.

Stanley, S.M. (1979) *Macroevolution: Pattern and Process*. San Francisco: W.H. Freeman.

Stensiö, E. (1968) 'The cyclostomes with special reference to the diphyletic origin of the Petromzontida and Myxinoidea' in *Current Problems in Lower Vertebrate Phylogeny. Nobel Symposium 4*, ed. T. Ørvig. Stockholm: Almquist & Wiskell.

Stevens, P.F. (1983) 'Report of the third annual Willi Hennig society meeting' in *Syst. Zool.*, 32:285–91.

Suppe, F. (1973) 'Facts and empirical truth' in *Can. J. Phil.*, 3:197–212.

Suppe, F. (1977) *The Structure of Scientific Theories*, 2nd edn. Urbana: Univ. of Illinois Press.

Swainson, W. (1834) *A Preliminary Discourse on the Study of Natural History*. Longman.

Tattershall, I. & Eldredge, N. (1977) 'Fact, theory and fantasy in human paleontology' in *Am. Scient.*, 65:204–11.

Thom, R. (1975) *Structural Stability and Morphogenesis. An Outline of a General Theory of Models*. Trans. D.H. Fowler. Reading, Mass.: W.A. Benjamin.

Thompson, D'Arcy W. (1942) *On Growth and Form*. Cambridge Univ. Press. 2nd edn.

Thompson, P. (1983) 'Tempo and mode in evolution: punctuated equilibria and the modern synthetic theory' in *Phil. Sci.*, 50:432–52.

Throckmorton, L.H. (1966) *The Relationships of the Endemic Hawaiian Drosophilidae*. Univ. Texas Publ., No 6615.

Troll, W. (1926) *Goethes Morphologische Schriften*. Jena: E. Diederichs Verlag.

Underwood, G. (1982) 'Parallel evolution in the context of character analysis' in *Zool. J. Linn. Soc.*, 74:245–66.

Valentine, E.R. (1982) *Conceptual Issues in Psychology*. London: Allen & Unwin.

Van Fraassen, B. (1980) *The Scientific Image*. Oxford: Clarendon Press.

Van Valen, L. (1978) 'Why not to be a cladist' in *Evolut. Theory*, 3:285-99.

Van Valen, L. (1982) 'Homology and causes' in *J. Morphol.*, 173:305-12.

Von Baer, K.E. (1828) *Entwicklungsgeschichte der Thiere: Beobachtung und Reflexion*. Konigsberg: Borntrager.

Voorzanger, B. & Van der Steen, W.J. (1982) 'New perspectives on the biogenetic law?' in *Syst. Zool.*, 31:202-5.

Waddington, C.H. (1957) *The Strategy of the Genes*. London: Allen & Unwin.

Wade, N. (1981) 'Dinosaur battle erupts in British Museum' in *Science*, 211:35-6.

Warburton, F.E. (1967) 'The purposes of classification' in *Syst. Zool.*, 16:241-5.

Watson, J.B. (1914) *Behaviour: Introduction to Comparative Psychology*. New York: Holt.

Watson, J.B. (1924) *Behaviourism*. New York: Holt.

Webster, G.C. & Goodwin, B.C. (1982) 'The origin of species: a structuralist approach' in *J. Soc. Biol. Struct.*, 5:15-47.

White, M.J.D. (1978) *Modes of Speciation*. San Francisco: W.H. Freeman.

Wiggins, D. (1980) *Sameness and Substance*. Oxford: Basil Blackwell.

Wiley, E.O. (1975) 'Karl Popper, systematics and classification – a reply to Walter Bock and other evolutionary taxonomists' in *Syst. Zool.*, 24:233-43.

Wiley, E.O. (1978) 'The evolutionary species concept reconsidered' in *Syst. Zool.*, 27:17-26.

Wiley, E.O. (1979) 'Cladograms and phylogenetic trees' in *Syst. Zool.*, 28:88-92.

Wiley, E.O. (1979) 'An annotated Linnaean hierarchy, with comments on natural taxa and competing systems' in *Syst. Zool.*, 28:308-37.

Wiley, E.O. (1980) 'Is the evolutionary species fiction? – a consideration of classes, individuals and historical entities' in *Syst. Zool.*, 29:76-80.

Wiley, E.O. (1981) *The Theory and Practice of Phylogenetic Systematics*, New York: John Wiley & Sons.

Wilson, E.O. (1965) 'A consistency test for phylogenies based upon contemporaneous species' in *Syst. Zool.*, 14:214-20.

Wittgenstein, L. (1953) *Philosophical Investigations*. Trans. G.E.M. Anscombe. Oxford: Basil Blackwell.

Woodger, J.H. (1945) 'On biological transformations' in *Essays on Growth and Form*, eds. W.E. Le Gros Clark & P.B. Medawar. Oxford: Clarendon Press.

Zangerl, R. (1948) 'The methods of comparative anatomy and its contribution to the study of evolution' in *Evolution*, 2:351-74.

AUTHOR INDEX

SUBJECT INDEX

absolute theory, *see* universals, absolute theory
Acanthopterygia, 32
adaptation, 44, 135, 163; 212n67
adaptive zones, 43, 65
additive model, 72, 80, 118; 200n109
 definition, 73
Aeluroidea, 62
agglomerative techniques, 201n23
allele, 61
Alligator, 54
allopatric speciation, *see* speciation, allopatric
Amniota, 79
Amphibia, 27, 79
anagenesis, 45–6, 59, 67, 174
anatomy, comparative, 26
ancestors
 common, 13–14, 16, 28–9, 43, 45, 50, 52, 55,
  69, 71, 74, 78, 82–3, 92, 121
 hypothetical, 71, 77
 see also relationships, ancestor–descendant;
  species, stem
Apidium, 49, 51
Araneae, 176
arbitrary, 83
 definition, 48
Archaeopteryx, 68
archetype, 18
Archosauria, 79
Arctoidea, 62
artificial classifications, *see* classifications, artificial
astronomical theory, 190n87
attributes, 14, 106, 114–15
 see also characters

Australopithecus, 49, 51
autapomorphy, 76
 see also synapomorphy
Aves, 27, 45–6, 53–4, 57, 59

Batrachomorpha, 79
Bauplan, 33, 61; 187n23; 191n98; 195n70
behaviourism, 204n82
'best-fit', 16, 117–18
bias, 5, 12, 23–31, 111
 elimination of, 6, 9–10, 119–27, 130, 139,
  142
 extensional, 23–4
 intensional, 23–4
binomial nomenclature, 116
biogeography, historical, 81
biological species concept, *see* species concepts,
  biological
blood relationships, *see* relationships, genealogical
  (blood)
Brachiopoda, 83
branching, 57
 order, 67, 135
 phylogenetic, 43
 point, 44, 56
 sequences, 4, 44, 46, 64, 71, 93, 99, 107, 182
 see also cladograms, phylogenetic trees
British Museum (Natural History), 185n3

Carnivora, 62
category, 14, 25, 120, 122–4
 definition, 90–1
 see also rank

238   *Subject Index*